ADVANCES
IN
ENVIRONMENTAL
PSYCHOLOGY
Volume 6
Exposure to Hazardous Substances:
Psychological Parameters

ADVANCES IN ENVIRONMENTAL PSYCHOLOGY

Volume 6

T.m ⌐Exposure to Hazardous Substances⌐: Psychological Parameters

Edited by **ALLEN H. LEBOVITS**
The Mount Sinai Medical Center
ANDREW BAUM
JEROME E. SINGER
Uniformed Services University of the Health Sciences

LAWRENCE ERLBAUM ASSOCIATES, PUBLISHERS
1986   Hillsdale, New Jersey   London

Lawrence Erlbaum Associates, Inc., Publishers
365 Broadway
Hillsdale, New Jersey 07642

Library of Congress Cataloging in Publication Data
Main entry under title:

Exposure to hazardous substances.

(Advances in environmental psychology ; v. 6)
Includes indexes.
1. Behavioral toxicology. 2. Hazardous substances--
Toxicology. I. Lebovits, Allen H. II. Baum, Andrew.
III. Singer, Jerome E. IV. Series. [DNLM: 1. Environ-
mental Exposure. 2. Environmental Pollutants--adverse
effects. 3. Stress, Psychological. WA 671 E96]
RA1224.E96 1986   616.9'8'0019   85-31163
ISBN 0-89859-860-5
Printed in the United States of America
10 9 8 7 6 5 4 3 2 1

Contents

List of Contributors

Kenneth M. Bachrach, *University of California, Los Angeles*

Andrew Baum, *Uniformed Services University of the Health Sciences*

Evelyn J. Bromet, *University of Pittsburgh School of Medicine*

Mary Byrne, *Mount Sinai School of Medicine*

Laura M. Davidson, *Uniformed Services University of the Health Sciences*

India Fleming, *Uniformed Services University of the Health Sciences*

Margaret S. Gibbs, *Fairleigh Dickinson University*

Martha M. Gisriel, *Uniformed Services University of the Health Sciences*

Elane M. Gutterman, *Mount Sinai School of Medicine*

Allen H. Lebovits, *Mount Sinai School of Medicine*

Adeline G. Levine, *State University of New York at Buffalo*

Jeffrey S. Markowitz, *Columbia University*

David K. Parkinson, *University of Pittsburgh School of Medicine*

Christopher M. Ryan, *University of Pittsburgh School of Medicine*

Jerome E. Singer, *Uniformed Services University of the Health Sciences*

Russell A. Stone, *State University of New York at Buffalo*

James J. Strain, *Mount Sinai School of Medicine*

Alex J. Zautra, *Arizona State University*

To Barry, Stacey, and Debra—
*May the world be a little
less hazardous when you
grow up*
—AHL

To Carrie
—AB

To Linda
—JES

Preface

Increasingly frequent environmental exposures to hazardous substances present mental health professionals with groups and at times communities of people, faced with high levels of psychological threat. As a result of an increasingly industrial and technological society, a new type of group cohort has emerged—individuals exposed to hazardous substances that present the possibility of immediate and chronic threats to their health and their families' health. Though epidemiologists, public health personnel, and medical researchers have established the medical sequelae to such exposure, little attention has been paid to mental health issues or to possible integrated psychophysiological consequences. The mental health profession may be called upon increasingly to assist in the psychological care of and attention to exposed individuals as well as in preparation of environmental impact estimates and in decisions about siting, operating, and clean-up. This book focuses on reactions to exposure to toxic substances as well as some predictors of response in groups faced with increased medical risk subsequent to some of the most common and hazardous toxic exposures found today: radiation, toxic waste, asbestos, lead, contaminated water, and toxic chemical fire and leak.

Despite the fact that the seven studies reported in this book differ methodologically in very important ways, such as time from and before exposure, instrumentation, and even social science orientation, several important common findings emerge. To obtain similarities in seven studies conducted completely independently of each other takes on added significance. In addition to psychological evaluation, the studies employed sociological, neuropsychological, behavioral, and physiological techniques to determine the nature of the relationship between exposure (or belief that one has been exposed) and mental health status.

With the exception of the two studies reporting on occupational exposures (Bromet, Ryan, and Parkinson; and Lebovits, Byrne, and Strain), several common findings emerge from the studies. First, all find elevated levels of psychological distress, both global and specific, among individuals subsequent to toxic exposure. Specific reactions found include higher than expected levels of demoralization, depression, and anxiety. The primary concern of exposed groups and groups who will be exposed is physical health, usually cited much more frequently than are other concerns such as financial loss. In some toxic exposure situations, the impact is apparently worse for those with young children. Length of time in community also appears to be an important variable, particularly when people become highly involved in the situation (e.g., by political activism).

Lack of information and conflicting messages regarding the exposure (or planned exposure) and its health hazard clearly exacerbate psychological reactions and contribute significantly to a lack of trust and confidence in the authorities and the government. Exposed individuals also report feelings of loss of control, helplessness, and powerlessness. These feelings may reflect the uncertainty that underlies much of the distress experienced after exposure. Because they are forms of response to loss of control, these feelings should reflect conditions that can exacerbate stress. Thus, the finding that helplessness or lack of control is associated with heightened distress should not be surprising.

The fact that nonoccupational exposure to toxic substances were associated with greater distress than were occupational exposures is an interesting collective implication of these studies. It may stem, in part, from the accidental nature of toxic exposure in residential or other nonoccupational settings. Clearly, exposure to hazards at Love Canal or Three Mile Island was not voluntary. Asbestos and lead workers were not exposed voluntarily either, but the nature of their exposure is clearly different. In addition, occupational exposure is more chronic, and increasing length of time could desensitize workers to dangers. Finally, sources of nonoccupational exposures typically are not sources of likelihood for all of those exposed, whereas in occupational situations, this is not so.

These studies all suggest the need for greater psychological attention to exposed individuals, even at the pre-exposure state when, for example, a toxic waste dump is being planned. Recognition of the pervasiveness of stressful reactions is an important first step in planning for a much needed intervention. Even in the occupational settings, intervention is needed to facilitate appropriate health promotive behaviors and to lessen the specific stressful reactions that emerge.

Allen H. Lebovits
Andrew Baum
Jerome Singer

OCCUPATIONAL EXPOSURES

1

The Case of Asbestos-Exposed Workers: A Psychological Evaluation

Allen H. Lebovits
Mary Byrne
James J. Strain
Mt. Sinai School of Medicine

Approximately 13.2 million workers were occupationally exposed to asbestos from 1940 to 1980 (Commercial Union Insurance Companies, 1982). Extensive exposure to asbestos occurred among shipyard workers during World War II when large quantities of asbestos were used, with an exposure of about 4.5 million people (Nicholson, Perkel, & Selikoff, 1982). This remarkable mineral's resistance to heat and chemicals and its great durability has led to its use in over 3,000 household and industrial products (Nicholson et al., 1982). Thirty million tons of asbestos are woven into the very fabric of our society—chemical plants, ships, power houses, buildings, and schools. In addition, the widespread use of asbestos in construction and other industries has resulted in the presence of substantial amounts of asbestos in existing structures with the potential for accidental or unknown exposure as these structures are remodeled, repaired, maintained, or demolished. Despite the approximately 20 Federal regulations which regulate environmental exposure to asbestos, large segments of the population continue to be exposed to asbestos.

The medical evidence that has accumulated during this century clearly indicates that inhalation of asbestos fibers leads to various severe and debilitating pulmonary diseases (Selikoff, Churg, & Hammond, 1964). Because the effects of inhalation of asbestos fibers are time and dose-related, certain of these diseases may develop only after the inhalation of substantial amounts of asbestos over a substantial period of time. Other diseases require only a comparatively small dose for their occurrence. The long latency periods that are characteristic of the actual diseases—that is, the period between initial exposure to asbestos and the

3

exhibition of the first clinical signs or symptoms of disease—indicate that the disabilities presented today may be reflective of exposures as long as 50 years earlier.

Longitudinal epidemiological studies of cohorts of asbestos insulation workers and shipyard workers have revealed that 19% of those with substantial exposure to asbestos die of lung cancer (Selikoff, Hammond, & Seidman, 1979). An additional 9% die of asbestosis, a chronic respiratory ailment, whereas 9% die of gastrointestinal cancer. Finally, 8% of those exposed to asbestos die of mesothelioma, a malignant tumor occurring in the chest or abdominal cavity lining that is exceptionally rare in the non-exposed population, and is a highly lethal disease. One of the most singular relationships that has been established between an oncogenic substance and a specific tumor and site exists between asbestos fiber and diffuse malignant mesothelioma (Selikoff, 1976; Selikoff et al., 1964; Sheers & Coles, 1980; Tagnon et al., 1980). The prognosis for patients who develop this disease is very poor: survival from onset of symptoms is generally less than 1 year (Chahinian et al., 1982; Harvey, Slevin, Ponder, Blackshaw, & Wrigley, 1984). The full impact of postwar experiences has not yet occurred because of the long latency period associated with mesothelioma (Tagnon et al., 1980).

Smoking by an asbestos worker carries with it a degree of risk substantially out of proportion to the risk associated with the same behavior by a member of the general population. Furthermore, cessation of smoking is followed by an actual decrease of risk of bronchogenic carcinoma among asbestos workers (Selikoff & Hammond, 1979; Selikoff, Hammond, & Churg, 1968). Smoking cessation is the single most preventive health measure that can be adapted by a smoker (Commercial Union Insurance Companies, 1982). The dynamics of smoking, however, are poorly understood and programs designed to eliminate the habit, although plentiful, have not been proven very effective. The need for increased risk awareness among blue-collar workers is particularly important. Blue-collar workers smoke more than any other type of employee (Sterling & Weinkam, 1978) and are likely to be exposed to occupational carcinogens. Though reduction or cessation of smoking does not apparently alter incidence of mesothelioma, it does appreciably affect the incidence of bronchogenic carcinoma. Asbestos-exposed individuals who smoke have a risk of lung cancer 53 times that of the general population of nonsmokers (Hammond, Selikoff, & Seidman, 1979). For asbestos-exposed smokers who smoke more than one pack per day the risk goes up to 87 times that of nonsmokers.

Awareness of the health hazards associated with asbestos exposure is becoming increasingly important as the pervasiveness of asbestos increases in the environment. Recent litigation with regard to responsibility concerning liability in cases of asbestos-related diseases has increased public consciousness of asbestos hazards (Brodeur, 1985).

RATIONALE FOR STUDY

A comprehensive psychological study has been undertaken to document the psychological adaptation and compliance behavior of individuals with well documented increased medical risk status due to asbestos exposure. The behavioral and psychological responses to the information that one is at increased risk for cancer has not been studied frequently, yet it bears careful evaluation in light of increasingly frequent environmental exposures to oncogenic substances. Data contributing to the understanding of health promotive behaviors, particularly smoking cessation among individuals exposed to toxic substances, especially in the workplace, is an urgent need and priority of preventive medicine. The few studies that evaluate groups at increased cancer risk show that being at increased risk appears to be a stressful state with a broad continuum of psychological responses.

Lynch and Krush (1968a, 1968b) evaluated individuals who were at increased cancer risk due to genetic reasons. Denial of risk was the apparent coping mechanism as evident from the observation that only 32% of one kindred and 53% of another kindred went to a physician for regular examinations. Nearly half the persons interviewed expressed feelings of apathy or fatalism regarding their increased risk for cancer, and tended to repress information.

Schwartz and Stewart (1977) interviewed the children of mothers who took diethylstilbestrol (DES) during their pregnancy. The children were at increased risk for developing genetic abnormalities and cancer, and daughters, in particular, were at increased risk for vaginal cancer. They studied the psychosocial and emotional effects on the daughters, and learned that the daughters reacted with varying degrees of anxiety, and were very upset about the iatrogenic nature of the contributory factor to their disease—the medication. They also found an incidental and nonprofessional communication of risk information. A DES Task Force (1979) reports that daughters frequently exhibit a variety of emotional problems and reactions such as anger, guilt, shame, fear of cancer, and problems in sexuality.

Michigan farmers exposed to polybrominated biphenyls (PBB) from contaminated animal feed in 1973 showed depressive symptoms, guilt, increased anxiety and concern, and emotional withdrawal (Brown & Nixon, 1979). They developed somatic defensive symptoms exemplified by many vague physical complaints. Other observers also found an increase in depression and anxiety in PBB-exposed Michigan farmers accompanied by many neurologic symptoms (Valciukas, Lilis, Anderson, Wolff, & Petrocci, 1979).

Levine (1982) and Gibbs (1983) have studied the Love Canal disaster from sociological and community perspectives. The events at Love Canal are described as a long-term crisis that produced a fear of developing illness, particularly cancer. Mistrust of authorities and lack of professional communication of information were major sources of stress.

Collins, Baum, and Singer (1983) and Fleming, Baum, Gisriel, and Gatchel (1982) have studied from a psychological and physiological vantage point, the effectiveness of different coping styles 2 years after the Three Mile Island (TMI) accident. Use of different coping strategies produced a selective susceptibility to stress, whereas high levels of social support mediated levels of stress such that TMI residents could cope more effectively. In a much simpler and less powerful study, Kasl, Chisholm, and Eskenazi (1981) evaluated workers at TMI 6 months after the accident and found them to have lower job satisfaction and greater uncertainty about their job future than a comparison group of nuclear workers. Hartsough and Savitsky (1984), however, have summarized the research on TMI and conclude that although stress levels in neighboring areas have increased sharply, particularly among mothers of young children, levels of stress were not traumatic and possibly not even chronic.

Li et al. (1983) evaluated the smoking behavior of 871 shipyard workers exposed to asbestos as part of an asbestos medical surveillance program. The threat of cancer did not appear to alter the daily activities of the workers: 43% of workers continued to smoke. There was a suppression of concern with a resigned acceptance of increased medical risk.

Groups of people who are exposed to hazardous substances may react with a great deal of uncertainty and psychological distress. Psychological attention to such individuals may become increasingly necessary. Scientific evaluations of increased medical risk groups may help elucidate common psychological responses.

STUDY DESCRIPTION

A cross-sectional study is presently being conducted to evaluate the behavioral and psychological responses to increased cancer risk status. Two groups of study participants have been entered onto study: (a) Asbestos workers, (pipecoverers or insulators), who were first exposed a minimum of 20 years ago and have continued their exposure into the last 5 years. (b) A comparison group of postal workers (letter carriers and mail handlers) who have worked for the postal service a minimum of 20 years and have no known occupational exposure to asbestos. All study participants were interviewed and given the following battery of standardized and specially-devised measures:

1. Project Questionnaire—This specially devised measure included the following areas of assessment: demographics, perception of disease susceptibility, acquisition of risk information, cognitive, and behavioral adaptations to risk information and attitudes.

2. Current and Past Psychopathology Scales (CAPPS)—Developed by Endicott and Spitzer (1972), this semi-structured interview consists of 173 ratings

by the interviewer of current and past psychopathology that are computer scored into 8 scales of current psychopathology and 18 scales of past psychopathology. After extensive training and the achievement of interrater reliability, every study participant was rated on a 6-point scale of psychopathological severity. It also has a computer-derived DSM II diagnosis and has been shown to be a reliable and valid measure (Spitzer & Endicott, 1969).

3. American Cancer Society Smoking Questionnaire—Based primarily on the Cancer Prevention Study Questionnaire of the American Cancer Society, this questionnaire has been used on over 1 million individuals including asbestos-exposed individuals (Hammond, 1966).

4. Global Assessment Scale (GAS)—The Global Assessment Scale (Endicott, Spitzer, Fleiss, & Cohen, 1976) is a single rating scale for evaluating the overall functioning of a person at a specific time period on a continuum of psychological health. The assessed time period is the past month, in order to correspond with the present section of the CAPPS.

5. Multidimensional Health Locus of Control Scale (MHLC)—The Multi-dimensional Health Locus of Control Scale (Wallston, Wallston, & DeVellis, 1978) is an 18-item self-report questionnaire assessing the personality dimension of internality–externality as it relates to health beliefs. The MHLC assesses feelings of internality and feelings of externality–powerful others and externality–chance as they relate to health and illness.

6. Impact of Stress Scale (ISS)—The Impact of Stress Scale is a 14-item self-report instrument (Horowitz et al. 1980) that is a revision of the Impact of Event Scale (Horowitz, Wilner, & Alvarez, 1979). The ISS has been adapted for use in the present study with slight modifications in its published form. It is designed to evaluate the stress of increased medical risk due to oncogenic exposure by classifying the respondents' thoughts and feelings regarding the risk into intrusive or avoidant responses.

Data is presented from an ongoing study for 111 asbestos workers and 48 postal workers.

Acquisition of Risk Information

Although the average year that asbestos workers participating in this study first entered the insulation business was 1954, it wasn't on the average until 1963 that they first acquired the risk information, that asbestos can be dangerous. More specific information, that asbestos exposure can lead to cancer, was learned on the average, 2 years later, in 1965. Finally, the most specific information, that smoking and asbestos exposure have a harmful relationship was not learned on the average until 1967, 13 years after starting in the business. Despite the fact that they didn't acquire risk information until they were well into their trade, they did learn about the asbestos dangers much before the general public. Postal workers recall learning about the dangers of asbestos in 1974, 11 years after the asbestos workers, and that asbestos exposure can lead to cancer in 1975, 10 years after the asbestos workers acquired the information. The synergistic relationship between smoking and asbestos exposure is not well known by non-asbestos workers; 94% of postal workers never learned of the harmful relationship.

The pattern of mode of acquisition of risk information for asbestos workers is dependent on type of information (Fig. 1.1). For the most general information, that asbestos can be dangerous to health, it is important to note that 39% of asbestos workers learned of their risk from nonprofessional sources (such as the public media and friends). As the information became more specific, professional involvement in the mode of acquisition became increasingly frequent ($\chi^2 = 25.59, p < .01$). The most specific information, the synergistic asbestos-smoking relationship, was acquired by 68% of the asbestos workers from professional sources. Only 3% of asbestos workers did not know of the synergistic relationship, indicative of a well-informed risk group. Postal workers, as expected, learned of asbestos dangers exclusively from nonprofessional sources.

There was an interaction between mode of acquisition of risk information and perceived initial concern (Table 1.1). Initial concern was reported more frequently by asbestos workers who learned of their risk from professional sources than by those who learned from nonprofessional sources. A stronger interaction effect was found with initial usage of masks: Those who learned from professional sources were significantly more likely to report initial usage of masks, than those who learned from nonprofessional sources ($\chi^2 = 5.50, p < .02$). Other behavioral effects however, such as smoking or physician visits were not related to mode of acquisition of risk information.

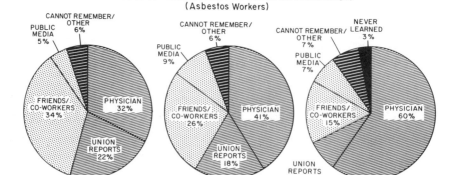

MODE OF ACQUSITION OF RISK INFORMATION
(Asbestos Workers)

DANGEROUS TO HEALTH   CAN LEAD TO CANCER   SMOKING AND ASBESTOS

$\chi^2 = 25.59, P < .01$

FIG. 1.1 Mode of acquisition of risk information (asbestos workers.)

TABLE 1.1
Effect of Mode of Acquisition of Risk Information

	Professional Sources	Non-Professional Sources
Report of Initial Concern	70%	56%
Report of Initial Non-Concern	30	44
Initial Usage of Masks	61%	39%
Initial Non-Usage of Masks*	39	61

$*X^2 = 5.50, p < .02.$

Disease Familiarity

Asbestos workers were surrounded by asbestos-related deaths. Seventy-eight percent of asbestos workers knew four or more coworkers who developed an asbestos related disease: lung cancer, mesothelioma, or asbestosis. They knew a median number of 11 such coworkers. Furthermore, 14% had a parent with one of those diseases. In contrast, 92% of postal workers did not know of any coworkers who developed one of these asbestos-related diseases ($\chi^2 = 107.72$, $p < .0001$), and knew a median number of less than one such coworker. Only 4% of postal workers had a parent with lung cancer, mesothelioma, or asbestosis.

The specific asbestos locals studied appeared to be a family enterprise. Of the asbestos workers, 78% had a relative in the insulation business compared to only 46% of postal workers who had a relative in the post office. Twenty-one percent of asbestos workers had their father in the trade in comparison with 2% of postal workers. Sixteen percent enrolled their own sons in the union, whereas only 6% of postal workers did so.

Disease Susceptibility

Every study participant was asked to rate, on a 4-point scale ranging from very unlikely to very likely, their chances for getting cancer and coronary heart disease (Fig. 1.2). Asbestos workers felt they were significantly more likely to develop cancer than did postal workers and more likely to develop cancer than coronary heart disease. Postal workers did not feel an increased susceptibility to cancer. Asbestos workers, therefore, were well aware of their risk of disease susceptibility.

Mental Health

An evaluation of current psychopathology (past month functioning) reveals no differences between both groups on any of the CAPPS scales (Fig. 1.3). Asbestos workers were not more impaired in reality testing, or more depressed or anxious,

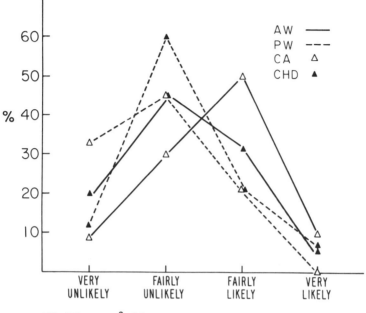

FIG. 1.2 Perception of disease susceptibility.

and manifested no differences on impulse control, or increased guilt. Of major interest is the fact that they did not somatize more and were not more concerned about their health. An evaluation of the past scales on the CAPPS reveals a generally similar lack of differences between the two groups. One of the unique features of the CAPPS is the computer-derived diagnostic system. For this study, the computer-derived DSM II diagnoses were converted to DSM III diagnoses based on the DSM III Manual guidelines. There were no significant differences in diagnostic classification between the two groups (Fig. 1.4).

With regard to the interviewer's ratings of global mental health, there were no significant differences between the two groups. The mean score for both groups was 75, which is in the range of "no more than slight impairment in functioning."

Neither group are frequent users of mental health services. Eighty-nine percent of asbestos workers have never been to see a mental health professional, compared to 85% of postal workers. Nine percent of asbestos workers and 6% of postal workers did have a consultation or brief period of treatment. Only 2% of asbestos workers were treated continuously or for several periods compared to 8% of postal workers.

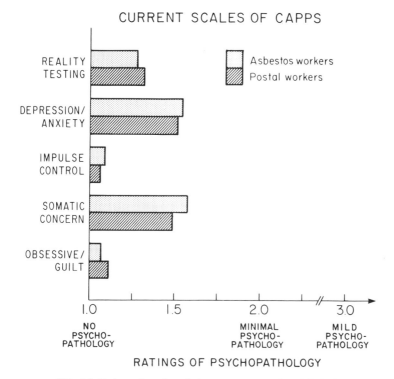

FIG. 1.3 Ratings of psychopathology on current scales of CAPPS.

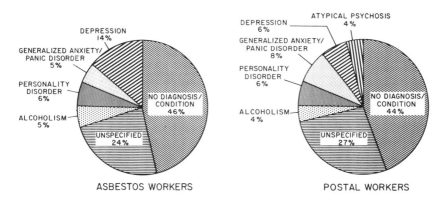

FIG. 1.4 Computer-derived diagnostic classifications.

Perceived Control of Health

There were no differences with regard to perceived control of health (Table 1.2). Asbestos workers did not feel less in control of their health than postal workers.

Behavioral Adaptations

Several behavioral responses to risk information were evaluated. The median percentage of time that asbestos workers used masks at work over all the years was only 9% whereas the percentage of workers who used a mask over 50% of the time was only 9%. Similarly, there was a lack of visits to a physician; 43% of the group never went for regular physical exams without symptoms.

Smoking is the single largest preventable public health problem in the United States today. For asbestos workers who smoke, it is the single most effective preventive health behavior that can be adopted. There were no significant differences between the smoking incidence of both groups. Thirty-two percent of asbestos workers continued to smoke, whereas another 45% smoked in the past. This is compared to 27% of postal workers who were current cigarette smokers, and another 46% who smoked in the past. Risk information had little impact on smoking behavior (Fig. 1.5). Of asbestos workers who had ever smoked, 19% had stopped before risk information, 9% had stopped within 1 year after risk information, 30% stopped after 1 year, and 42% continued to smoke. More asbestos worker smokers did not stop subsequent to risk information than did stop.

A closer evaluation of the group at highest increased risk, the asbestos worker smokers, showed that present smokers perceived their health to be controlled externally by chance factors significantly more than past and never smokers did (Table 1.3). The perception that health is not one's own responsibility, but rather is largely contingent on chance, luck, or fate may provide a convenient rationalization for smoking. This is consistent with previous findings showing present smokers to be more externally oriented in their health perceptions (Kaplan & Cowles, 1978).

The Impact of Stress Scale evaluates how increased medical risk status is experienced: intrusively or in an avoidant manner (which the scale authors feel

TABLE 1.2
Perceived Control of Health

MHLC	Asbestos Workers	Postal Workers
Internality	27.5	27.5
Externality-Chance	15.9	18.2
Externality-Powerful Others	20.6	22.3

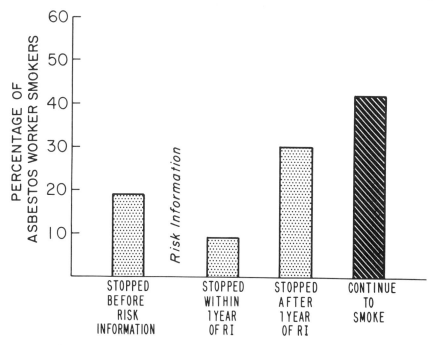

FIG. 1.5 Smoking cessation and risk information.

TABLE 1.3
Smoking and Perceived Control of Health

MHLC Scales	Asbestos Workers		
	Present Smokers	Past Smokers	Never Smokers
Internality	27.5	26.7	28.9
Externality–Chance*	18.1	15.7	13.8
Externality–Powerful Others	21.7	20.6	19.3

*$F = 3.3$, $p < .05$.

is very similar to denial). Increased risk status was significantly less intrusive for present smokers than for never smokers (Table 1.4). There was also a trend for present smokers to have higher avoidant scores. Present smokers appear to have been thinking of their risk less by denying it more.

CONCLUSIONS

Risk information communicated from a professional source tends to be associated with an initial increase in concern and usage of masks as compared to risk information communicated from a nonprofessional source. This is in consonance with social psychology theory regarding the effectiveness of the authoritative communication of risk information (Janis, 1974). Modern prevention efforts designed to screen populations for risk factors associated with illnesses such as lung disease, hypertension, or genetic diseases, have as a major component the dissemination of risk information and the adaptability of health promotive behaviors. Recognition of the importance of the communication of risk information, particularly who the communicator is, is essential. The effectiveness of this information in motivating alteration of risk factors such as smoking cessation and promoting preventive health behaviors such as participation in medical surveillance and treatment programs is important. Notification of workers at high risk is an emerging public health problem that has significant scientific, legal, and economic implications (Schulte & Ringen, 1984).

Asbestos workers did not appear to be more psychologically distressed than the comparison group. Furthermore, risk information did not appear to appreciably affect health behaviors such as smoking cessation, or long-term usage of masks. The findings strongly suggest that despite a strong cognizance of their risk, asbestos workers did not adapt health promotive behaviors and did not experience excessive levels of psychological distress. Asbestos worker smoker–nonsmoker differences centered on perceptions of control of health and the "intrusiveness" of medical risk status.

Preventive health behaviors are multidetermined and it may be simplistic to believe that any single construct such as perception of control of health will predict much of the variance in individual health behaviors. A multivariate

TABLE 1.4
Impact of Stress Scale & Smoking

	Present Smokers	Past Smokers	Never Smokers
Intrusion Score*	6.9	9.4	11.2
Avoidant Score	10.6	8.6	8.7

*$F = 3.39, p < .04.$

evaluation may best predict multidetermined behaviors. The study of the effect of chronic stress on complex health behaviors and cognitions outside the psychological laboratory setting is difficult. Issues of reliability of observation and control of extraneous and confounding variables become germane. However, the ecological validity of this study permits the observation of "real" behaviors and cognitions in the industrial worker's natural setting. Such a setting may allow for a more valid response.

The asbestos workers who have been evaluated are constantly reminded of their risk from friends and coworkers dying, the public media, medical surveillance exams, union reports, and research projects. Despite their risk awareness, noncompliant behaviors are abundant: continuing in the trade, smoking, and lack of usage of masks. As Horowitz et al. (1980) and Janis (1974) have previously shown, the typical reaction to risk information is denial, a defense that reduces subjective distress. The lack of excess psychological distress observed in this study is in consonance with the workers' behaviors. The fact that the asbestos insulation trade may be a family type of business, a way of life almost, that happens to pay very well, may help support the denial system.

The lack of observed psychological distress along with nonadaptive behaviors may indicate that denial is a common coping mechanism characteristic of chronic occupational exposures. Intervention is clearly needed to combat denial and enhance the adaptive use of risk information to support health promotive behavior.

ACKNOWLEDGMENTS

This research was supported by National Institute of Environmental Health Sciences grant #5R01 ES02578. Grateful appreciation is acknowledged to Drs. Irving Selikoff, James F. Holland, A. Philippe Chahinian, and Jimmie Holland, for their efforts in this research. We also wish to thank Dr. John Nee, Dr. John Thornton, and Jonine Bernstein for their computer assistance, and Marlaine Larosiliere for her very able preparation of this manuscript.

REFERENCES

Brodeur, P. (1985, June, July). Annals of law—The asbestos industry on trial. *New Yorker, 61.*

Brown, G. G., & Nixon, R. (1979). Exposure to polybrominated biphenyls: Some effects on personality and cognitive functioning. *Journal of the American Medical Association, 242,* 523–527.

Chahinian, A. P., Pajak, T. F., Holland, J. F., Norton, L., Ambinder, R. M., & Mandel, E. M. (1982). Diffuse malignant mesothelioma—Prospective evaluation of 69 patients. *Annals of Internal Medicine, 96,* 746–755.

Collins, D. L., Baum, A., & Singer, J. E. (1983). Coping with chronic stress at Three Mile Island: Psychological and biochemical evidence. *Health Psychology, 2,* 149–166.

Commercial Union Insurance Companies (1982). *Asbestos, smoking and disease: The scientific evidence.* Boston, MA: Author.

DES Task Force Summary Report. DHEW Publication No. (NIH) 79–1688.

Endicott, J., & Spitzer, R. L. (1972). Current and Past Psychopathology Scales (CAPPS). *Archives of General Psychiatry, 27,* 678–687.

Endicott, J. E., Spitzer, R. L., Fleiss, J. L., & Cohen, J. (1976). The Global Assessment Scale— A procedure for measuring overall severity of psychiatric disturbance. *Archives of General Psychiatry, 33,* 766–771.

Fleming, R., Baum, A., Gisriel, M. M., & Gatchel, R. J. (1982). Mediating influences of social support on stress at Three Mile Island. *Journal of Human Stress, 8,* 14–22.

Gibbs, L. M. (1983). Community response to an emergency situation: Psychological destruction and the Love Canal. *American Journal of Community Psychology, 11,* 116–125.

Hammond, E. C. (1966). Smoking in relation to the death rates of one million men and women. In *Epidemiological Study of Cancer and Other Chronic Diseases* (pp. 127–204), Monograph 19. National Cancer Institute.

Hammond, E. C., Selikoff, I. J., & Seidman, H. (1979). Asbestos exposure, cigarette smoking, and death rates. *Annals of the New York Academy of Science, 330,* 473–490.

Hartsough, D. M., & Savitsky, J. C. (1984). Three Mile Island: Psychology and environmental policy at a crossroads. *American Psychologist, 39,* 1113–1122.

Harvey, V. J., Slevin, M. L., Ponder, B. A., Blackshaw, A. J., & Wrigley, P. F. (1984). Chemotherapy of diffuse malignant mesothelioma. *Cancer, 54,* 961–964.

Horowitz, M., Hulley, S., Alvarez, W., Billings, J., Benfari, R., Blair, S., Borhani, N., & Simon, N. (1980). News of risk for early heart disease as a stressful event. *Psychosomatic Medicine, 42,* 37–46.

Horowitz, M. J., Wilner, N. J., & Alvarez, W. (1979). Impact of Event Scale: A measure of subjective stress. *Psychosomatic Medicine, 42,* 209–218.

Janis, I. L. (1974). Vigilance and decision making in personal crises. In G. V. Coelho, D. A. Hamburg, & J. E. Adams (Eds.), *Coping and adaptation* (pp. 139–176). New York: Basic Books.

Kaplan, G. D., & Cowles, A. (1978). Health locus of control and health value in the prediction of smoking reduction. *Health Education Monographs, 6,* 129–137.

Kasl, S. V., Chisholm, R. F., & Eskenazi, B. (1981). Part I: Perceptions and evaluations, behavioral responses, and work-related attitudes and feelings. *American Journal of Public Health, 71,* 472–483.

Levine, A. G. (1982). *Love Canal: Science, politics, and people.* Lexington, MA: D.C. Heath.

Li, V. C., Kim, Y. J., Terry, P. B., Cathie, J. C., Roter, D., Emmett, E. A., Harvey, A., & Permutt, S. (1983). Behavioral, attitudinal, and physiologic characteristics of smoking and nonsmoking asbestos-exposed shipyard workers. *Journal of Occupational Medicine, 25,* 864–870.

Lynch, H., & Krush, A. J. (1968a). Delay: A deterrent to cancer detection. *Archives of Environmental Health, 17,* 204–209.

Lynch, H., & Krush, A. J. (1968b). Heredity, emotions and cancer control. *Postgraduate Medicine, 43,* 134–138.

Nicholson, W. J., Perkel, G., & Selikoff, I. J. (1982). Occupational exposure to asbestos: Population at risk and projected mortality—1980–2030. *American Journal of Industrial Medicine, 3,* 259–311.

Schulte, P. A., & Ringen, K. (1984). Notification of workers at high risk: An emerging public health problem. *American Journal of Public Health, 74,* 485–491.

Schwartz, R. W., & Stewart, N. B. (1977). Psychological effects of diethylstilbestrol exposure. *Journal of the American Medical Association, 237,* 252–254.

Selikoff, I. J. (1976). Lung cancer and mesothelioma during prospective surveillance of 1249 asbestos insulation workers, 1963–1974. *Annals of the New York Academy of Science, 271,* 448–456.

Selikoff, I. J. (1977). Cancer risk of asbestos exposure. In H. H. Hiatt, J. D. Watson, & J. A. Winsten (Eds.), *Origins of human cancer, Cold Spring Harbor Conferences on cell proliferation* (Vol. 4, pp. 1765–1784). Cold Spring Harbor, NY: Cold Spring Harbor Laboratory.

Selikoff, I. J., Churg, J., & Hammond, E. C. (1964). Asbestos exposure and neoplasia. *Journal of the American Medical Association, 188,* 22–26.

Selikoff, I. J., & Hammond, E. C. (1979). Asbestos and smoking (editorial). *Journal of the American Medical Assocation, 242,* 458.

Selikoff, I. J., Hammond, E., & Churg, J. (1968). Asbestos exposure, smoking and neoplasia. *Journal of the American Medical Association, 204,* 106–112.

Selikoff, I. J., Hammond, E. C., & Seidman, H. (1979). Mortality experience of insulation workers in the United States and Canada, 1943–1976. *Annals of the New York Academy of Science, 330,* 91–116.

Sheers, G., & Coles, R. M. (1980). Mesothelioma risks in a naval dockyard. *Archives of Environmental Health, 35,* 276–282.

Spitzer, R. L., & Endicott, J. (1969). DIAGNO II: Further developments in a computer program for psychiatric diagnosis. *American Journal of Psychiatry, 125,* 12–21.

Sterling, T. S., & Weinkam, J. J. (1978). Smoking patterns by occupation, industry, sex and race. *Archives of Environmental Health, 33,* 313–317.

Tagnon, I., Blot, W. J., Stroube, R. B., Day, N. E., Morris, L. E., Peace, B. P., & Fraumeni, J. F. (1980). Mesothelioma associated with the shipbuilding industry in coastal Virginia. *Cancer Research, 40,* 3875–3879.

Valciukas, J. A., Lilis, R., Anderson, H. A., Wolff, M. S., & Petrocci, M. (1979). The neurotoxicity of polybrominated biphenyls: Results of a medical field survey. *Annals of the New York Academy of Science, 320,* 337–367.

Wallston, K. A., Wallston, B. S., & DeVellis, R. (1978). Development of the Multidimensional Health Locus of Control (MHLC) Scales. *Health Education Monographs, 6,* 161–170.

2 Psychosocial Correlates of Occupational Lead Exposure

Evelyn J. Bromet
Christopher M. Ryan
David K. Parkinson
University of Pittsburgh School of Medicine

Inorganic lead is one of the most common sources of exposure in industry worldwide. More than 1 million tons of lead are consumed yearly by American industries, and potential exposure to lead and its compounds occurs in at least 100 occupations. It has long been known that lead intoxication can affect the integrity of both the central and peripheral nervous systems, with acute encephalopathy sometimes resulting from extreme lead poisoning. Recognizing the potential health hazards, the Occupational Safety and Health Administration issued a new lead standard that mandates the maximum allowable blood and air lead values for lead-exposed workers, use of protective equipment, and other safety procedures for the lead industry.

Although the deleterious neuropsychiatric effects of lead poisoning have repeatedly been demonstrated (e.g., Browder, Joselow, & Louria, 1973), the relationship between low-level lead exposure and central nervous system (CNS) dysfunction in adults has not been firmly established. Some investigators have reported statistically reliable differences between exposed workers and controls on measures of personality and anxiety (Haenninen, Hernberg, Mantere, Vesanto, & Jalkanen, 1978; Repko, Morgan, & Nicholson, 1974; Spivey et al., 1979) or on neuropsychological assessments of visuomotor and cognitive functioning (Grandjean, Arnvig, & Beckmann, 1978; Hogstedt, Hane, Agrell, & Bodin, 1983; Repko, Corum, Jones, & Garcia, 1978; Valciukas et al., 1980) while others found no effects on psychosocial (Repko et al., 1978) or neuropsychological variables (Haenninen et al., 1978). Statistically significant dose-response relationships also have been reported for psychological (Baker, Feldman, & White, 1981; Lilis et al., 1977) and neuropsychological (Baker et al., 1984;

19

Haenninen et al., 1978; Valciukas et al., 1980) indices, although non-significant effects have been reported on both psychosocial (Haenninen et al., 1978; Repko et al., 1978; Repko et al., 1974; Spivey et al., 1979) and neuropsychological (Grandjean et al., 1978; Repko et al., 1978) measures as well. Haenninen et al. (1978) suggested two methodological weaknesses to explain these inconsistencies: the reliance on current exposure measures rather than cumulative exposure indices; and inadequate sensitivity of the psychological tests for detecting subtle effects.

In addition, sampling inadequacies and varying data collection procedures may have contributed to the inconsistent findings across studies. With two exceptions (Baker et al., 1984; Spivey et al., 1979), the samples were composed of volunteers, with the "non-exposed" subjects included without being prescreened for prior lead exposure. Control subjects often differed from the lead-exposed workers on demographic and premorbid characteristics (e.g., age, education, alcohol consumption) that are known correlates of performance on the dependent measures (Grandjean et al., 1978; Lilis et al., 1977; Repko et al., 1978). The most extreme examples of inappropriate sampling comes from a study in which the neuropsychological test results of 90 volunteers from a lead smelter in Los Angeles were compared with those of Wisconsin farmers and Virginia papermill workers (Valciukas et al., 1978), and with Michigan residents exposed to polybrominated biphenyls (Valciukas et al., 1980).

With respect to variations in field procedures that could influence performance on the psychological tests, only one study (Spivey et al., 1979) used examiners who were blind to the exposure status of the workers. The examinations themselves were conducted in various settings (i.e., at an office, medical facility, or the plant) and at varying times (before, during, or after work). In a recent study, Baker et al. (1984) interviewed on plant premises during working hours. Although this procedure facilitated participation, it introduced bias arising from non-blind examiners, distraction from noise, and worker fatigue. In some cases, there was incomplete data because subjects returned to work before finishing the testing.

Recently, it has been suggested that the interaction between neurotoxic exposure and occupational stress may be implicated in the etiology of adverse health outcomes in workers (Ashford, 1976). Some evidence to support this synergy hypothesis with respect to physical health effects has been reported (House, Wells, Landerman, McMichael, & Kaplan, 1979). It is conceivable that the inconsistencies across the occupational lead literature may be due in part to a failure to consider more complex models that reflect interactions between low-level exposure and degree of work-related stress.

To examine potential neuropsychiatric effects of occupational lead exposure, we conducted an epidemiologic study designed to mitigate the methodological difficulties just described. Our study evaluated randomly selected exposed and non-exposed factory workers from plants in the same geographic area. The non-exposed group was extensively prescreened for prior exposure to lead as well as other toxic substances. A comprehensive test battery incorporating well-known neuropsychological and psychological tests was administered outside the work-place by examiners blind to the workers' exposure status. Complete lead exposure records for each worker were obtained from the lead battery plants. Finally, psychiatric and substance abuse histories were obtained, and blood alcohol and urine drug screens were performed at the time of testing. The present study addresses three fundamental questions: Is the psychosocial and neuropsychological functioning of lead workers significantly poorer than that of controls? Among lead workers, do higher levels of exposure result in impairments in neuropsychiatric functioning? Is there a synergistic effect of lead exposure and occupational stress on psychosocial functioning?

MATERIALS AND METHODS

Sample

The present study was based on 288 lead-exposed workers drawn from three battery plants and 181 non-exposed controls from a heavy industrial plant who had no known neurotoxic exposures. All four plants were located in eastern Pennsylvania and organized by the United Steel Workers of America. Eligibility requirements included current employment for at least 1 year, being male, 18–60 years of age, a native English speaker and Caucasian. These criteria reflected the dominant characteristics of the workforce in the four plants.

The samples were drawn from company listings containing name, address, telephone number, age, and for workers in the battery plants, their most recent blood lead value. In order to examine a range of cumulative exposure levels, we increased the proportion of older workers with longer term exposure by stratifying workers in each plant into three age groups (18–25, 26–30, 31–60), and drawing proportionate random samples of 10%, 30%, and 60%, respectively. Refusal rates were 32% among exposed and 33% among control workers. Comparison between participants and refusers from the lead battery plants showed no differences in their most recent blood lead values. In one lead plant, and in the control plant, refusers were somewhat older than participants. In addition we subsequently excluded three control workers because their current blood levels exceeded 35 mg/100 ml and two lead plant workers because the company records showed that their blood lead levels had never exceeded 35 mg/100 ml.

Potential controls were prescreened by telephone for evidence of prior lead exposure. A questionnaire was developed that delineated 25 jobs containing lead exposure; workers were excluded if they held these jobs for at least 6 months in the last 10 years. Those not excluded in this way were asked about hobbies involving soldering and excluded if they soldered at least twice a week for the last 2 months. Approximately one third of the prescreened workers were found ineligible because of prior lead exposure.

Demographically, the exposed workers were significantly younger ($\bar{x}$ = 35.22 ± 10.32 vs. 39.48 ± 11.05; $p < .001$), and less educated ($\bar{x}$ = 11.07 ± 1.72 vs. 11.52 ± 1.64; $p < .01$) than the controls. As a consequence, the lead plant workers also were employed for fewer years (10.57 ± 7.87 vs. 15.12 ± 11.62; $p < .001$) and earned less income than the controls (17% vs. 58% were earning > \$30,000/year; $p < .001$). Psychiatric and substance abuse histories were determined using the Diagnostic Interview Schedule (Robins, Helzer, Croughan, & Ratcliff, 1981), a standard epidemiologic instrument, and no differences between exposed and control workers were found prior to employment at the study plants. Workers also were screened for serious head injury, neurological, renal, and hepatic disorders. Thus in the analysis, only age and education were selected as covariates.

Data Collection

Two interviews were conducted. The first was a home interview that focused on psychosocial parameters. In the present study, seven measures serve as dependent variables. The first three are direct measures of psychological status, namely Trait Anger and Trait Anxiety, measured by the 35-item Spielberger scale (Spielberger, Gorsuch, & Lushine, 1970); and Average Number of Drinks per Day, a summary measure that combines quantity–frequency information for the past month (Cahalan, Cisin, & Crossley, 1969). The remaining four focus on potential indirect effects of disturbed psychological state, namely, stressful interpersonal relationships and number of work-related accidents in the past 10 years. The interpersonal stressors also were evaluated because of anecdotal reports from the workers that lead exposure made them angry and irritable. These measures, derived by comparing results of factor analyses conducted separately for each plant, were:

> *Conflict with Friends* (4 items): sum of 3 items dealing with frequency of respondents' expressing anger toward friends and 1 item on friends' getting angry with respondent.
> *Conflict with Relatives* (2 items): sum of 1 item focused on respondents' having bad arguments with relatives and 1 item on relatives' not talking to respondent because of his temper.
> *Conflict with Spouse* (3 items): sum of 2 items dealing with help-seeking for marital problems and 1 item on time spent thinking about marital problems (Pearlin, Lieberman, Menaghan, & Mullan, 1981).

The second interview involved assessments of occupational stress and neuropsychological performance, and took place within 2 weeks of the first session in an office rented for the study. To avoid fatigue, workers were tested on weekends or before work. Workers also were instructed to refrain from drug or alcohol use within 12 hours of testing and their self-reports were confirmed by alcohol and drug screens, all of which were negative. Ten neuropsychological variables covering five general domains are included in the analysis as dependent variables.

General Intelligence (3 tests)

Information: Subjects were asked 29 general information questions from this subtest of the Wechsler Adult Intelligence Scale, Revised (WAIS-R). Raw scores from this, and from all other WAIS-R subtests, were age-scaled (mean = 10; standard deviation = 3), according to the WAIS-R standardization norms (Wechsler, 1981).

Picture Completion: Twenty pictures of objects and scenes were presented, and subjects were asked to identify what crucial part was missing (standard WAIS-R scoring).

Similarities: Subjects were given 14 word pairs and asked to tell how each pair was alike (standard WAIS-R scoring).

Visuoconstructional Ability (1 test)

Block Design: Subjects were asked to reproduce a series of increasingly difficult geometric designs using red and white blocks (standard WAIS-R scoring).

Attention/Psychomotor Integration (3 tests)

Digit Symbol Substitution: Subjects were asked to rapidly replace a series of randomly arrayed numbers with symbols according to a specified code (standard WAIS-R scoring).

Digit Span: Subjects were first asked to repeat exactly sets of numbers of increasing length and then to repeat other sets in reverse order (WAIS-R scoring).

Benton Visual Retention Test, Form F: A geometric design was presented for 5 seconds and then removed. Subjects were then presented with four designs and asked to select the original. Fifteen such trials were administered (Benton, 1965).

Visuoperceptual Ability (1 test)

Embedded Figures Test: Subjects were asked to find a simple design embedded in one of four complex patterns. Ten such trials were administered (Kapur, & Butters, 1977).

Motor Speed/Manual Dexterity (1 test)

Grooved Pegboard Test—Dominant and Non-Dominant Hands: Time taken by subject to insert 25 notched pegs into a board containing 25 matching holes. Two measures are derived, the first for the dominant hand and the second for the non-dominant hand (Lezak, 1983).

During the second interview, blood samples were taken on all participants and analyzed by an OSHA-certified laboratory. Blood lead levels (PbB) were determined using atomic absorption spectrophotometry. In the present analysis, PbB and zinc protoporphyrin levels (ZPP) were used as indicators of current exposure. Three measures of prior exposure, based on records of blood lead values provided by the lead battery plants, were calculated: time-weighted average (TWA); peak exposure level (PK); and proportion of blood lead values greater than 60 mg/100 ml (PR60), that is, higher than the 1981 OSHA-mandated threshold value. It should be noted that calculations were not done on workers with fewer than four recorded blood lead values. The median number of values was 41; there were 40 men whose blood lead levels had been determined on more than 60 occasions. Finally, length of employment in the battery plant was included as an exposure indicator. Table 2.1 presents the intercorrelations among the six exposure variables, as well as their means and standard deviations. The three measures based on prior lead levels are strongly correlated with one another. Length of employment is unrelated to the current indices and moderately correlated with prior exposure levels.

Finally, three occupational stress measures were derived for the analysis of interaction effects: Conflict at Work, a four-item scale measuring hostility with coworkers and foreman; Job Satisfaction, a four-item scale derived from the work of House et al. (1979); and Control, a two-item scale indicating feelings of lack of control over work-related matters.

Training

Comprehensive training programs were developed for both the psychosocial interviewers and the neuropsychological examiners. Training sessions took place over a period of 6 weeks, and included didactic seminars, practice interviews with volunteers from a local plant, and reliability checks. During the field phase, the trainers observed test sessions for the first five subjects and then randomly for the duration.

TABLE 2.1
Intercorrelations Among Lead Exposure Measures for 288 Lead-Exposed Workers[a]

	$\bar{x}$	s.d.	ZPP	TWA	PK	PR60	LOE
PbB	40.01	13.17	.46	.58	.35	.44	.01
ZPP	91.63	67.32	—	.41	.23	.33	.10
TWA	48.83	11.90		—	.75	.87	.32
PK	78.78	28.78			—	.73	.36
PR60	0.23	0.22				—	.25
LOE	10.57	7.87					—

[a]All coefficients, except LOE vs. PbB and LOE vs. ZPP, significant at $p < .001$.

Analysis

Comparisons between exposed and control workers were conducted using analysis of covariance, controlling for age and education. Dose-response relationships were determined by partial correlation coefficients that also controlled for age and education. The correlations among the neuropsychological and psychosocial measures were relatively modest; therefore we did not employ a multivariate analytic strategy. Finally, multiple regression was used to test for synergistic effects.

RESULTS

Between-Group Comparisons

Descriptive statistics for exposed and control workers on the psychosocial and neuropsychological dependent variables are presented in Table 2.2. The exposed workers were not significantly different from the controls on most of the psychosocial and neuropsychological measures. In one instance (Digit Symbol Substitution) the lead workers performed better than the controls. On the other hand, in two areas, the lead workers were significantly worse than the controls, namely, the number of work-related accidents and performance on the Grooved Peg Board.

Because lead is a cumulative toxin, we next examined whether the subset of workers with longer term exposure would be more discrepant compared to their non-exposed peers than was found for the sample as a whole. We identified 54 lead workers 40+ years of age with 13+ years of exposure (median for the lead workers) and 84 controls of similar age and length of employment. Similar results were obtained. Number of Work Related Accidents significantly differentiated between the groups, with the absolute difference in mean number of accidents being greater for the older lead workers $(.73 \pm .11)$ versus older controls $(.28 \pm .09)$; $(p < .01)$ than for the sample as a whole. In addition, older lead workers did more poorly than controls on two neuropsychological tests, the Grooved Pegboard using the non-dominant hand $(88.27 \pm 2.03$ vs. 79.79 ± 1.61; $p < .01)$ and Information $(8.98 \pm .28$ vs. $9.92 \pm .22$; $p < .01)$.

Dose-Response Effects.   Table 2.3 presents the partial correlation coefficients for the six lead measures and each of the psychosocial and neuropsychological variables for the lead exposed workers. Although length of exposure was unrelated to the psychosocial measures, the other lead exposure measures, particularly those reflecting cumulative exposure, were related in the predicted direction to conflict with friends, and to a lesser extent with Number of Work Related Accidents. There were very few statistically significant correlations between the

Table 2.2
Differences between Exposed and Non-Exposed Workers in
Psychosocial and Neuropsychological Functioning: Analysis of Co-
Variance Adjusting for Age and Education

	Exposed		Controls		
Psychosocial Variables	$\bar{x}^a$	sd	$\bar{x}^a$	sd	F
Conflict with Friends	9.56	2.43	9.30	2.47	ns
Conflict with Relatives	1.39	.87	1.40	.88	ns
Marital Conflict[b]	4.69	1.54	4.56	1.67	ns
Average Number of Drinks	1.68	2.42	1.68	2.43	ns
Trait Anger	28.83	6.77	27.99	6.86	ns
Trait Anxiety	37.41	8.01	36.42	8.11	ns
Number of Accidents	.55	.81	.20	.82	18.43***
Neuropsychological Variables					
Information	9.00	2.14	9.24	2.16	ns
Picture Completion	8.48	1.99	8.68	2.01	ns
Similarities	8.47	2.27	8.48	2.29	ns
Block Design	10.12	2.38	10.24	2.41	ns
Digit Symbol Substitution	9.30	1.98	8.78	2.00	7.12
Digit Span	9.04	1.96	9.05	2.12	ns
Benton Visual Retention Test	12.32	1.58	12.55	1.60	ns
Embedded Figures Test	7.76	1.74	7.72	1.76	ns
Grooved Pegboard Test—Dominant Hand	74.47	11.42	71.66	11.56	6.32**
Grooved Pegboard Test—Non-Dominant Hand	79.04	12.75	75.36	12.89	8.74**

[a]Adjusted means.
[b]N's = 236 (exposed) and 151 (non-exposed)
*p < .05; **p < .01; ***p < .001.

exposure and the neuropsychological measures, and they accounted for only 1%–2% of the variance.

When the analysis was restricted to older lead workers employed for 13 + years, the absolute size of many of the dose-response relationships increased. However, fewer of the coefficients reached statistical significance because of the smaller sample size. For example, among the psychosocial variables, the correlations with TWA increased to .24 ($p < .05$) for Trait Anger, .20 ($p < .09$) for Trait Anxiety, and .22 ($p < .06$) for Accidents. Among the neuropsychological variables, no clear-cut pattern was evident.

Table 2.3
Relationship between Exposure and Psychosocial Measures among
Lead Battery Plant Workers: Partial Correlations Controlling for
Education and Age

	PbB	ZPP	TWA	PK	PR60	LOE
Psychosocial						
Conflict with Friends	.13*	.06	.22***	.14**	.18***	.03
Conflict with Relatives	−.02	−.00	.02	.07	.05	−.00
Marital Conflict	−.04	−.08	−.12*	−.07	−.13*	.04
Average Number of Drinks	.03	−.07	−.02	−.02	.05	.03
Trait Anger	.05	−.01	.11*	.05	.07	.05
Trait Anxiety	.03	−.05	−.00	−.03	.01	.04
Number of Accidents	.02	−.02	.11*	.08	.15***	−.01
Neuropsychological						
Information	−.00	.02	.05	.07	.03	.03
Picture Completion	.09	.08	.05	.05	.09	−.07
Similarities	−.01	.03	−.02	.02	−.05	−.03
Block Design	−.08	−.00	−.12*	−.07	−.10*	−.06
Digit Symbol Substitution	−.04	.10	−.02	−.01	−.01	−.06
Digit Span	−.04	.07	−.01	.08	−.03	.00
Benton Visual Retention Test	−.03	−.01	−.00	−.03	.02	−.02
Embedded Figures Test	−.08	−.02	−.11*	−.07	−.07	−.04
Grooved Pegboard Test—Dominant Hand	.02	−.02	.01	.05	.04	.00
Grooved Pegboard Test—Non-Dominant Hand	.02	.01	.09	.10*	.07	.14*

$*p < .05; **p < .01; ***p < .001.$

Interaction of Exposure and Stress. Finally, a series of regression analyses was conducted to identify possible significant interaction effects. Separate analyses were conducted to consider interactions between TWA and PbB with each of the three stress variables. In these analyses, age and education were first entered as a block, followed by a job stress and exposure block; the interaction term was entered last. The dependent variables selected for these analyses were Anger, Anxiety, Conflict with Friends, and performance on the two Grooved Peg Board tests. Thus, a total of 30 analyses were conducted (3 stress measures × 2 exposure variables × 5 dependent variables). In only one instance was the change in R^2 statistically significant. Thus there was no evidence for the presence of a unique synergistic effect of exposure and stress on psychosocial functioning in the present sample. The amount of variance explained for the psychosocial measures ranged from 4%–11%. For the Grooved Peg Board tests, the percentage of variance explained was higher (15%–23%), due almost entirely to effects of age and education.

DISCUSSION

The present study examined the relationships between occupational lead exposure, occupational stress, and neuropsychiatric functioning in a large sample of workers exposed from 1 to 39 years. Findings based on comprehensive neuropsychological and psychosocial assessments revealed few significant effects. Exposed workers did not differ on most variables from non-exposed controls, nor were measures of prior and current lead exposure related to neuropsychological test scores. On the other hand, lead exposure was somewhat related to one psychosocial parameter, namely, conflict in interpersonal relationships. The same pattern of results was obtained for a subsample of older workers who had been exposed to lead for a prolonged period of time. Finally, no support was found for an interaction effect involving lead and occupational stress.

Previous findings on the relationship between neuropsychological performance and occupational lead exposure have been inconsistent. When we reviewed the evidence supporting significant effects of low-level lead exposure, we found that most papers reported neuropsychological test scores that were within the normal range. Moreover, in some cases there appeared to be a multiple comparison problem because the number of statistically significant associations was counterbalanced by an even larger number of nonsignificant effects. Thus if we ignore the fact that the methodology was inadequate in many cases, it is still likely that the "significant" findings have been over-interpreted.

It is interesting to note that similarly inconsistent results have been obtained in studies of the peripheral nervous system. Only two studies of nerve conduction velocities used an epidemiologic approach (Baloh et al., 1980; Nielsen, Nielsen, Kirkby, & Gyntelberg, 1982), and both produced negative results. In particular, Nielsen et al. (1982) examined all 95 employees working at a lead smelter for 9+ years and found no evidence of peripheral neuropathy. Thus, our findings of essentially normal central nervous system functioning, as indexed by neuropsychological test results, matches that of similarly designed studies of the peripheral nervous system.

With respect to psychosocial adjustment, the relationship between lead exposure and interpersonal conflict is consistent with Spivey et al.'s (1979) findings of aggressive behavior and hostility in lead-exposed workers. These results also are consonant with our clinical impression regarding high levels of hostility and irritability in our lead-exposed patients. They also confirm anecdotal descriptions by spouses of personality changes occurring prior to documented blood lead elevations. However, these significant correlations need to be interpreted cautiously, because the majority of coefficients were non-significant and the magnitude of the effect was modest.

The present study focused on employed workers. Because we did not have access to workers who left the plants, we were concerned about the potential effects of selection bias. One could argue that those who developed neuropsy-chiatric disturbances from lead exposure were more likely to quit their jobs. The only relevant data we could review to evaluate this potential bias were Worker Compensation cases, and we found no records that indicated the presence of neurological damage. Another source of selection bias that could have influenced our findings were factors associated with seeking employment at a lead battery plant rather than a plant with no neurotoxic exposures. Thus, we examined a set of preemployment variables, such as demographic factors and psychiatric and substance abuse histories, and controlled in the analysis for those that dif-ferentiated between exposed and control workers. A final source of sampling bias was the tendency for refusers from one of the lead plants and the control plant to be somewhat older than participants from those two plants. However, we have no reason to assume that the older men who participated differed from those who refused in ways that might have influenced the findings. Although we cannot eliminate the possibility that selection bias influenced our results, we suspect that its effect is relatively inconsequential.

The failure to document interactive effects of lead and occupational stress is consistent with the House et al. (1979) findings of synergism only occurring in relation to physical health rather than mental health outcomes. Thus, although the hypothesis remains plausible, we were not able to provide confirmation in this particular population.

In reviewing the overall pattern of results, we have formulated the following tentative conclusions about mental health sequelae of lead exposure. Clearly, very high levels of lead exposure often lead to somatic and CNS problems in adult workers. On the other hand, lower exposure levels do not appear to affect cognitive functioning although they may affect psychosocial functioning. Because there were few significant between-group differences, the correlations between lead exposure and psychosocial functioning variables must be interpreted cau-tiously. These tentative conclusions do not imply that lower level exposure has no biological impact on workers. Certainly biochemical changes, such as increased ZPP and inhibition of ALADH, are still present. Animal studies (Silbergeld & Lamon, 1980) suggest that CNS dysfunction may be related to elevations of ALA, and its interaction with gamma-aminobutyric acid, and it is possible that the correlations we reported between lead and psychosocial functioning are asso-ciated with these well recognized biochemical changes.

Major improvements in the degree of lead exposure have taken place since 1978 when the new lead standard was implemented. Blood lead records, as well as air lead records, document that our sample of lead workers is no longer exposed to the high levels prevailing prior to 1978. The fact that we detected no significant neuropsychological differences among workers with very high prior exposure histories (18 of the 54 older workers had peak blood lead levels of more than 120 ug/100 ml) suggests that even if acute effects existed, reversal of these effects may well have occurred (Silbergeld, 1983). In summary, our investigation provided no evidence that adverse neuropsychiatric effects occurred as a result of either low-level lead exposure or the interaction between exposure and occupational stress.

ACKNOWLEDGMENTS

This research was supported in part by National Institute of Mental Health Grant No. MH36221.

We wish to thank the research staff of the Psychiatric Epidemiology Program for assisting with all phases of the study and local union officers and staff, District 7, United Steel Workers of America for their help in implementing the field work.

REFERENCES

Ashford, N. (1976). Crisis in the workplace: Occupational disease and injury. Cambridge, MA: MIT Press.

Baker, E. L., Feldman, R. G., & White, R. (1981). Neuropsychiatric effects of occupational lead exposure—Preliminary findings of a longitudinal study. In C. Perris, G. Struwe, & B. Jansson (Eds.), Biological Psychiatry (pp. 114–121). Amsterdam: Elsevier North-Holland Biomedical Press.

Baker, E. L., Feldman, R. G., White, R. A., Harley, J. P., Niles, C. A., Dinse, G. E., & Berkey, C. S. (1984). Occupational lead neurotoxicity: A behavioural and electrophysiological evaluation. Study design and year one results. British Journal of Industrial Medicine, 41, 352–361.

Baloh, R. W., Spivey, G. H., Brown, C. P., Morgan, D., Campion, D. S., Browdy, B. L., Valentine, J. L., & Gonick, H. C. (1980). Subclinical effects of chronic increased lead absorption—A prospective study. II. Results of baseline neurologic testing. Journal of Occupational Medicine, 21, 490–496.

Benton, A. L. (1965). Manuel du test de retention visuelle: applications cliniques et experimentales (2nd ed.). Paris: Centre de Psychologie Appliquee.

Browder, A. A., Joselow, M. M., & Louria, D. B. (1973). The problem of lead poisoning. Medicine, 52, 121–139.

Cahalan, D., Cisin, I., & Crossley, H. (1969). American drinking practices. New Brunswick, NJ: Rutgers Center of Alcohol Studies.

Grandjean, P., Arnvig, E., & Beckmann, J. (1978). Psychological dysfunctions in lead-exposed workers. Scandinavian Journal of Work, Environment and Health, 4, 295–303.

Haenninen, H., Hernberg, S., Mantere, P., Vesanto, R., & Jalkanen, M. (1978). Psychological performance of subjects with low exposure to lead. Journal of Occupational Medicine, 20, 683–689.

Hogstedt, C., Hane, M., Agrell, A., & Bodin, L. (1983). Neuropsychological test results and symptoms among workers with well-defined long-term exposure to lead. British Journal of Industrial Medicine, 40, 99–105.

House, J. S., Wells, J. A., Landerman, L. R., McMichael, A. J., & Kaplan, B. H. (1979). Occupational stress and health among factory workers. Journal of Health and Social Behavior, 20, 139–160.

Kapur, N., & Butters, N. (1977). An analysis of visuoperceptive deficits in alcoholic Korsakoffs and long-term alcoholics. *Journal of Studies on Alcohol, 33,* 2025–2035.

Lezak, M. (1983). *Neuropsychological assessment* (2nd ed.). New York: Oxford University Press.

Lilis, R., Fischbein, A., Diamond, S., Anderson, H. A., Selikoff, I. J., Blumberg, W. E., & Eisinger, J. (1977). Lead effects among secondary lead smelter workers with blood lead levels below 80 ug/100 ml. *Archives of Environmental Health, 32,* 256–266.

Nielsen, C. J., Nielsen, V. K., Kirkby, H., & Gyntelberg, F. (1982). Absence of peripheral neuropathy in long-term lead-exposed subjects. *Acta Neurologica Scandinavica, 65,* 241–247.

Pearlin, L., Lieberman, M., Menaghan, E., & Mullan, J. (1981). The stress process. *Journal of Health and Social Behavior, 22,* 337–356.

Repko, J. D., Corum, C. R., Jones, P. D., & Garcia, Jr., L. S. (1978). *The effects of inorganic lead on behavioral and neurologic function* (DHEW [NIOSH] Publication No. 78–128). Washington, DC: U. S. Government Printing Office.

Repko, J. D., Morgan, Jr., B. B., & Nicholson, J. (1974). *Behavioral effects of occupational exposure to lead* (DHEW [NIOSH] Publication No. 75–164). Washington, DC: U. S. Government Printing Office.

Robins, L. N., Helzer, J. E., Croughan, J., & Ratcliff, K. S. (1981). National Institute of Mental Health Diagnostic Interview Schedule. *Archives of General Psychiatry, 38,* 381–389.

Silbergeld, E. K. (1983). Experimental studies of lead neurotoxicity: Implications for mechanisms, dose-response, and reversibility. In M. Rutter & R. R. Jones (Eds.), *Lead versus health: Sources and effects of low level lead exposure* (pp. 191-216). Chichester, England: Wiley.

Silbergeld, E. K., & Lamon, J. M. (1980). The role of altered haem synthesis in the neurotoxicity of lead. *Journal of Occupational Medicine, 25,* 680–684.

Spielberger, C. D., Gorsuch, R. L., & Lushine, R. E. (1970). *Manual for the State-Trait Anxiety Inventory.* Palo Alto: Consulting Psychologists Press, Inc.

Spivey, G. H., Brown, C. P., Baloh, R. W., Campion, D. S., Valentine, J. L., Massey, Jr., F. J., Browdy, B. L., & Culver, B. D. (1979). Subclinical effects of chronic increased lead absorption— A prospective study. I. Study design and analysis of symptoms. *Journal of Occupational Medicine, 21,* 423–429.

Valciukas, J. A., Lilis, R., Eisinger, J., Blumberg, W. E., Fischbein, A., & Selikoff, I. J. (1978). Behavioral indicators of lead neurotoxicity: Results of a clinical field survey. *International Archives of Occupational and Environmental Health, 41,* 217–236.

Valciukas, J. A., Lilis, R., Singer, R., Fischbein, A., Anderson, H. A., & Glickman L. (1980). Lead exposure and behavioral changes: Comparisons of four occupational groups with different levels of lead absorption. *American Journal of Industrial Medicine, 1,* 421–426.

Wechsler, D. (1981). *Manual for the WAIS-R.* New York: The Psychological Corp.

II NON-OCCUPATIONAL EXPOSURES

3 Toxic Exposure and Chronic Stress at Three Mile Island

Laura M. Davidson, Andrew Baum, India Fleming, and
Martha M. Gisriel
Uniformed Services University of the Health Sciences

The accident at the Three Mile Island (TMI) nuclear power station occurred in
March 1979. Although the original mishap may not have been as acutely stressful
and overwhelming as other cataclysmic events (e.g., tornadoes, floods, aircraft
accidents), sources of threat appear to have remained for people living near the
damaged reactor. The continuation of threat after an event like this is unusual,
and one possible reason for the persistence of problems at TMI may be the toxic
nature of the accident. Because the power plant generated nuclear power, the
accident involved the possibility of radioactive exposure among nearby populations. But, because of conflicting information given by authorities, many residents report that they do not know how much, if any, radiation they were
exposed to or what kinds of effects they might reasonably expect to observe.
Thus, many fear the future consequences of possible radiation exposure that may
have already occurred, worrying that they or their children might develop cancer
some time in the future. Also, as work on the damaged reactor has continued
and still progresses, the threat of radiation exposure has remained. As long as
radioactive waste from the accident remains, this threat may persist for many
area residents. Therefore, the toxic nature of the accident may have made it a
source of continued stress for TMI area residents.

In this chapter we discuss data obtained from people living near TMI that
show that 28 months after the accident, TMI area residents continued to exhibit
higher levels of stress than did people in control areas less affected by the disaster.
In addition, some possible reasons for the persistent stress are examined, and
potential mediators of chronic stress are considered.

STRESS

Stress is the process by which an organism is threatened by environmental events known as stressors. During stress, organisms must resist, adapt, or otherwise find ways of coping with environmental demands. If coping is not successful or if the stress is intense or persistent, negative consequences are likely to follow. The negative consequences may reflect psychological or physiological changes and thus play a role in the development of disease states including cardiovascular diseases like coronary heart disease and hypertension, immune system dysfunction, peptic ulcers, anxiety, and depression (e.g., Amkraut & Solomon, 1977; Ross & Glomset, 1976; Selye, 1956; Weiner, 1977).

Traditionally, theories of stress have followed one of two approaches using either physiological or psychological perspectives to conceptualize the process. Selye's (1936) work on stress is perhaps the best-known physiological perspective. He found that animals responded to a variety of noxious stimuli (e.g., hormones, heat, irritants) with the same changes. These changes consisted of a triad of responses including enlargement of the adrenal glands, shrinkage of the thymus, and ulceration of the gastrointestinal tract. According to Selye this response was nonspecific; all noxious stimuli produced the same triad of responses. In order to explain the progression from stimulus presentation to organ damage, Selye further proposed the three-stage general adaptation syndrome (GAS). The first stage, alarm, is characterized by recognition of threat and preparation to resist it. After the organism becomes aware of the stressor, it readies itself to respond by increasing vital functions such as adrenal activity, respiration, and cardiovascular activity. Once the various systems have been mobilized, the organism enters the stage of resistance that continues until reserves have been depleted or the stressor is overcome. If the stressor persists or coping abilities are low, the stage of exhaustion is entered. This stage is associated with diseases of adaptation (e.g., cardiovascular diseases, immune diseases). In the most extreme cases death occurs.

Other theorists have emphasized psychological factors in their conceptualizations of stress (e.g., Lazarus, 1966). According to this general perspective, appraisals are a crucial factor in stress responding. The emphasis is not on response, as it was in Selye's model, but on the interaction between the organism and the environment. Only when events are appraised as threatening will stress responding ensue (Lazarus, 1966). The importance of the appraisal process has been suggested by studies that examined response to similar stressors under different conditions. One compared a group of conscious dying patients with a group of comatose dying patients (Symington, Currie, Curran, & Davidson, 1955). These researchers examined the adrenal glands from both groups at autopsy and found that only the conscious dying group exhibited hypertrophy. A series of studies conducted by Lazarus and his colleagues provide further support for the role of psychological factors in stress responding (e.g., Lazarus, Opton,

Nomikos, & Rankin, 1965). In one set of experiments they encouraged either threatening or benign appraisals before subjects viewed anxiety-provoking or threatening films. Prior to viewing the films, several different explanations were given. One film consisted of a series of woodshop accidents involving injuries or mutilations. Subjects were told either that the film had been staged (the accidents were not real), that it would be used to prevent future accidents, or were given no explanation. Results suggested that stress responding depended on the cover story given. If subjects were given no explanation, stress responding ensued, but when they were given explanations that allowed them to discount or intellectualize (i.e., it was staged, it would be used to prevent future accidents), stress responding did not occur.

According to Lazarus (1966), once a situation is appraised as threatening, secondary appraisals are made. Secondary appraisals involve estimations of the benefits of coping in different ways. The two basic forms of coping are direct action and palliative responses. In direct action coping, the organism focuses its attention toward altering its relationship with the stressor. Leaving the stressful environment or manipulating the stressor are examples of this coping approach. When an organism chooses palliative forms of coping, it attempts to accommodate to the stressor by reducing or managing its emotional response to the stressor. Taking drugs, using alcohol, relaxing, or otherwise minimizing a stressor's emotional impact are examples of this form of coping.

Our research has suggested that one way of coping with stress may be by increasing perceptions of personal control (Baum, Fleming, & Singer, 1983). Depending on the situation, control may be maintained by using different forms of coping; but by increasing perceptions of control, adaptation to stress may be facilitated. For example, Rodin, Solomon, and Metcalf (1978) found that control reduced crowding stress, and Staub, Tursky, and Schwartz (1971) found that control could reduce perceptions of discomfort caused by physical shock. Likewise, perceptions of control seem to have played a role in mediating the development of stress symptoms in TMI area residents.

THE SITUATION AT TMI

It is commonly accepted that the accident at TMI as well as the 2-week emergency period that followed it were stressful. During the accident, extremely high temperatures were generated in the core of the reactor causing equipment to melt and fuse. This caused radioactive contamination of the reactor building as well as the containment building that surrounds it. Also, radioactive krypton gas was trapped in the containment building and leaked sporadically into the atmosphere until it was deliberately released into the environment about 15 months after the accident. More recently, radioactive water was removed from the floor of the reactor building, but the contaminated core still remains.

During the 2-week emergency period that followed the accident, evacuation was recommended for certain groups including pregnant women and mothers with preschool children. Actual numbers of evacuees reflected more widespread fear. It has been estimated that 30%–50% of the population living within 5 miles of the plant and up to one third of the population living between 10 and 15 miles of the plant evacuated the area (Flynn & Chalmers, 1980; Kraybill, Buckley, & Zmuda, 1979). The average length of time that people stayed away was 5 days, and the average distance traveled was 100 miles (Houts, Miller, Tokuhata, & Ham, 1980). Reasons for leaving included perceived danger (Flynn, 1979), fear of forced evacuation, and confusion surrounding the accident (Flynn, 1979; Houts et al., 1980).

Immediately following the accident, evidence suggested increased levels of stress among TMI residents. For example, the use of alcohol, tobacco, sleeping pills, and tranquilizers increased following the accident (Houts et al., 1980). Also, physicians reported more somatic complaints and higher blood pressure (Behavioral Medicine Special Report, 1979), and anxiety and depression were heightened among mothers of preschool children (Bromet, 1980). Other reported symptoms of stress following the accident including demoralization and emotional upset (Dohrenwend, Dohrenwend, Kasl, & Warheit, 1979).

Although some researchers reported improvement in stress symptoms with time, most agree that sources of stress still remained for TMI residents. At least 25% of the population reported that they still felt threatened by TMI 5 months after the accident (Flynn & Chalmers, 1980). However, few studies have examined the chronic effects of living near TMI. Bromet (1980) studied mothers with preschoolers immediately after the accident and 9 months later. At both times she found more anxiety and depression among the TMI mothers than in a control population. Houts et al. (1980) also examined more long-term consequences of the accident. A telephone survey was used to evaluate the effects of the accident in July 1979 and again in 1980, 9 months following the accident. Their results paralleled Bromet's in that they continued to find elevated levels of distress 9 months after the accident. In this study, symptoms of distress included loss of appetite, overeating, trouble sleeping, feeling shaky, trouble thinking clearly, irritability, and anger.

We began studying TMI area residents 15 months after the accident. At that time officials planned to begin the controlled venting of the radioactive krypton gas that had been trapped in the containment building. This venting procedure provided the opportunity to study the anticipatory effects of the venting, the effects of the actual venting, and the post-venting effects. We found higher levels of stress among TMI subjects as compared to three separate control groups. Although results indicated that stress levels were highest among TMI subjects just prior to the venting, they continued to be elevated relative to controls (Baum, Fleming, & Singer, 1982). Although stress levels seemed to decrease somewhat during the venting procedure, they were elevated again 6 months following the completion of the venting (Collins, Baum, & Singer, 1983).

Subject Population

There were 103 subjects in this part of the study. Thirty-six subjects were selected in a quasi-random fashion from neighborhoods within 5 miles of TMI. Another 27 were selected from Frederick, Maryland, a town at least 80 miles away from TMI. Two other control groups were selected. One group of 15 subjects lived in Dickerson, Maryland, within 5 miles of a coal-fired power plant. The final group of 25 subjects lived within 5 miles of the undamaged nuclear reactor at Oyster Creek, New Jersey. In all areas, streets were randomly selected, and every third house was approached. Response rates averaged 75% at all sites.

Stress Measures

The fact that physiological and psychological changes and processes are involved in stress suggests that it is a "whole-body" response. In order to measure this stress response, a multilevel research strategy was employed. This method involves the simultaneous assessment of psychological, behavioral, and physiological indices of stress.

Self-report measures of psychological stress were collected that included a global index of symptom reporting, anxiety, fear, depression, somatic complaints, concentration problems, interpersonal problems, anger, suspiciousness, and alienation. Self-reported symptoms of stress were measured with the SCL-90R (Derogatis, 1977). Subjects were asked how often they had been bothered by each of the 90 symptoms during the previous 2 weeks. Responses were made on 5-point scales (0–4) ranging from "not at all" bothered by the symptoms to "extremely" bothered by the symptoms.

Two behavioral measures of stress also were obtained. One, called the Towers of Hanoi, was used to measure persistence. The task employs a board on which there are three wooden pegs. On one of the pegs is a group of six rectangular pieces of wood placed so that they form a pyramid. Subjects are asked to move the pyramid to another peg one piece of wood at a time. At no point may they put a longer piece of wood on top of a shorter one. The solution is not very difficult, but it requires persistence if a subject is not familiar with it. The second behavioral measure, proofreading, was used to assess ability to concentrate. Subjects were given 5 minutes to read and locate errors in a seven-page passage from Jacobs' *The Death and Life of Great American Cities*. Subjects were asked to circle misspellings, typographical errors, punctuation errors, and grammatical errors. This task was similar to the one used by Glass and Singer (1972) in their studies of noise and stress.

Physiological measures of stress were obtained by assaying subjects' urine samples for the catecholamines epinephrine and norepinephrine. Fifteen-hour samples were collected (between 6 p.m. one evening and 9 a.m. the next morning). The urine was measured and a small sample frozen so that it could be assayed at a later time using COMT radioenzymatic procedures (Durrett & Ziegler, 1980).

EVIDENCE OF STRESS

Previous study of these groups, 17 and 22 months after the accident, indicated that TMI area residents exhibited more stress by reporting more distress on the self-report measures (e.g., Baum, Gatchel, & Schaeffer, 1983). This was again the case. TMI residents reported more symptoms in general, as well as more intense somatic distress, depression, and anxiety. They also reported more intense symptoms of alienation, suspiciousness, fear, and problems with concentration (see Table 3.1).

Behavioral measures also indicated that TMI residents experienced more difficulty with persistence and concentration than control area subjects. Overall, TMI residents spent less time working on the tower task and made fewer moves. Performance on the proofreading passage also indicated a deficit on the part of TMI subjects. Overall, TMI residents found fewer of the inserted errors than did Frederick, Dickerson, or Oyster Creek subjects (see Table 3.2).

For the biochemical indices of stress, TMI area residents exhibited higher levels of epinephrine and norepinephrine than did all control area subjects (see Table 3.3).

Stress and Control

Clearly, psychological, behavioral, and biochemical indices of stress indicated that TMI area residents continued to show stress 28 months after the accident. Measures of stress at this time were comparable to those collected 17 and 22 months after the accident. One possible explanation for the continuing stress is

TABLE 3.1
Symptom Reporting ($\bar{X}$ Levels) by TMI and Control Subjects

	TMI	Frederick	Dickerson	Oyster Creek
Total Symptoms Reported	32.7[a]	16.5[b]	21.6[b]	15.6[b]
Somatic Distress	.60[a]	.30[b]	.37[b]	.20[b]
Depression	.67[a]	.35[b]	.41[b]	.28[b]
Anxiety	.58[a]	.23[b]	.26[b]	.17[b]

Note: Means sharing superscripts are not significantly different from one another.

TABLE 3.2
Mean Task Performance by TMI and Control Subjects

	TMI	Frederick	Dickerson	Oyster Creek
Proofreading Errors (%)	52.0[a]	70.2[b]	68.6[b]	65.6[b]
Towers of Hanoi Task				
Total Moves	100.6[a]	215.8[b]	168.4[b]	175.9[b]
Total Time (in seconds)	689.4[a]	1234.0[b]	1110.0[b]	904.2[b]

Note: Means sharing superscripts are not significantly different from one another.

TABLE 3.3
Mean Catecholamine Levels in TMI and Control Subjects

	TMI	Frederick	Dickerson	Oyster Creek
Norepinephrine (ng/hr)	1776.8[a]	773.4[b]	806.4[b]	574.6[b]
Epinephrine (ng/hr)	659.5[a]	298.6[b]	185.4[b]	178.8[b]

Note: Means sharing superscripts are not significantly different from one another.

that the nature of the accident might pose different threats than many other stressors. In particular, the event was technological, and we believe that the effects of technological disasters may be different from those associated with natural events (Baum, Fleming, & Davidson, 1983). This distinction may have implications for people's perceptions of control after the occurrence of a disaster. Further, the event posed the threat of radiation exposure, and the involvement of toxins such as radiation may dramatically alter the nature of an event.

In some ways the accident at TMI was unprecedented. It was a major nuclear accident, the cause of which was not an act of nature like a tornado or a blizzard, but rather was a reflection of human error. Over the years we have created a vast network of technology in order to improve our lives. Unfortunately, our ability to regulate this technology has not always kept pace with our advances. For example, it still seems impossible to predict when technology will break down. Emergency procedures are often not perfected. This seemed to be the case at TMI. Although people expect to encounter storms or other types of natural events once in a while, technological mishaps are not anticipated in the same way. It is possible that these differences may influence people's perceptions of control. Because perceptions of control influence the stress response, it is a potentially important variable to consider as a mediator of the stress experienced by the residents at TMI.

Technological disasters may threaten control in a different way than do natural disasters. This may be because they represent a loss of control over something that at one time seemed to be under control. Natural forces are not ever under control, so destructive storms and the like do not reflect a loss of control but rather a lack of control. Research has explored many aspects of control, but when considering stress, one of the most important findings is that the impact of an event is influenced by the perceived controllability of the event (Glass & Singer, 1972). If one believes that one has control over an aversive condition, even if control is not real or is never used, subsequent stress will be reduced. In addition, loss of control may be a stressor (Wortman & Brehm, 1975). Therefore, the effects and aftereffects of a technological disaster may be greater than those of a natural disaster. It is for these reasons that we felt that control might be related to the stress experienced by our TMI subjects.

We have some evidence that suggests that TMI residents perceive themselves as having less control over their surroundings than do comparison subjects. Three questions were used to assess different aspects of perceived control. Subjects were asked to indicate how often they experienced feelings of helplessness, how often they felt that it did not matter what they did when given choices, and how often they felt that they did not care whether they did one thing or another. In fact, TMI subjects reported more feelings of helplessness than did all of the comparison groups and more often reported feeling that it didn't matter what they did when given choices. Similarly, they felt more frequently that they did not care whether they did things one way or another. These data seem to indicate that residents at TMI were experiencing more control-related problems than were subjects from the other groups.

To follow this up, we examined the relationship between control and stress. It was hypothesized that subjects experiencing control-related problems also would exhibit more symptoms of stress. TMI area subjects who were experiencing control-related problems should show the greatest amount of stress. In order to test this hypothesis, subjects were placed in groups based on their responses to the three questions that were used to assess perceived control. Those reporting above the median levels were considered to have more perceived control, whereas those below the median were classified as having less perceived control. These groupings were then crossed with residents (TMI vs. all of the other control groups combined), and analyses were performed using measures of stress as dependent variables.

Self-reported symptoms of stress indicated that TMI area residents who reported lower perceived control reported the most symptoms (see Table 3.4). They also indicated experiencing the most somatic distress and depression.

The behavioral data showed similar results. Both the Towers of Hanoi and the proofreading task showed patterns that were similar to the self-report measures (see Table 3.5). Findings for persistence, measured as both time spent on the tasks and number of attempts made, showed that the TMI subjects who were low on the control variable also were the ones to spend less time on the task; and they made fewer attempts to solve it. Similarly, these people found fewer

TABLE 3.4
Symptom Reporting and Expectations for Control

	$\bar{X}$ number of symptoms reported		$\bar{X}$ rating of somatic distress		$\bar{X}$ rating of depression symptoms	
	High	Low	High	Low	High	Low
TMI	35.11[a]	45.61[b]	.38[b]	.76[a]	.74[b]	.97[a]
Control	10.11[c]	12.8[c]	.22[b]	.34[b]	.17[c]	.23[c]

Note: Means sharing superscripts are not significantly different from one another.

TABLE 3.5
Performance on the Proofreading and Towers of Hanoi Tasks by
Subjects' Expectations for Control (Above and Below Median Levels)

	% Proofreading errors found		Total moves		Total time	
	High	Low	High	Low	High	Low
TMI	62.26[b]	45.61[a]	172.16[b]	89.65[a]	998.00[b]	598.22[a]
Control	70.83[b]	61.63[b]	197.02[b]	117.40[b]	1112.0[b]	829.27[b]

Note: Means sharing superscripts are not significantly different from one another.

TABLE 3.6
Catecholamine Levels by Subjects' Expectations for Control (Above
and Below the Mean)

	$\bar{X}$ levels of urinary norepinephrine (ng/hr)		$\bar{X}$ levels of urinary epinephrine (ng/hr)	
	High	Low	High	Low
TMI	819.44[b]	2072.64[a]	286.77[b]	763.55[a]
Control	623.13[b]	1185.15[c]	190.30[b]	451.45[c]

Note: Means sharing superscripts are not significantly different from one another.

of the errors on the proofreading task. Urinary norepinephrine and epinephrine levels yielded a similar pattern; TMI subjects reporting less perceived control exhibited higher levels of arousal than did any of the other subjects (see Table 3.6).

Although control-related problems probably do not account for all of the difficulties at TMI, these data indicate that they may play a role in the stress response. Another factor that may contribute to chronic stress is uncertainty. Several aspects of the situation at TMI could generate uncertainty for area residents. For example, threats at the crippled plant have remained despite clean-up activities. There is a more or less constant threat of exposure to radiation as work on the crippled reactor has progressed. First, radioactive gas was trapped in the containment building that was gradually released 15–16 months after the accident. Only recently, the radioactive water was removed from the floor of the containment building, and still the reactor core is yet to be removed. Throughout this time of clean-up activities the possibility of further radiation exposure has remained. Therefore, the consequences of the event may have persisted because of the nuclear nature of the accident. In addition, some of the residents fear that they have been exposed to radiation, and not only were they powerless to do anything to prevent the exposure, they are currently unable to do anything to counteract the possible long-term consequences of that exposure. Because the effects of radiation exposure may not materialize for years, people who believe that they have been exposed to radiation may be uncertain about their future health. These aspects of the accident may have generated a great deal of uncertainty among area residents. In fact, not only did TMI area residents express more concern about possible radiation leaks, they also reported more concern about the threat to their personal health and more concern about the threat to the health of their family members than have residents of the control areas.

CONCLUSIONS

TMI still continues to pose problems for many of the area residents. We have been able to document a persistent stress reponse psychologically, behaviorally, and biochemically more than 2 years after the accident. These problems may be partly explained by the nature of the accident. Technological disasters are not planned for in the same way as natural disasters and hence may threaten control more directly. We have collected data that suggest that control-related problems exacerbate stress responses or are otherwise associated with stress among TMI area residents. TMI also was unique in that it posed the additional threat of radiation exposure. Many of the residents fear the future consequences of this exposure and are uncertain of the future costs of that exposure. This uncertainty may have exacerbated the loss of control experienced by the subjects. All of these aspects of the situation may have been responsible, in part, for the continuing stress experienced by the people living near the reactor.

In order to conceptually replicate the results presented in this chapter, we began studying a group of people living near a toxic dump site. The site chosen for study was identified in 1982 by the Environmental Protection Agency (EPA) as one of the 10 most hazardous sites in the country. Our data from residents living near the site suggest that these people are exhibiting psychological, behavioral, and physiological symptoms of chronic stress (Fleming & Baum, 1985–1986). Also, other data from this sample parallel the stress data from TMI area residents in another way. Among residents living near the landfill, stress levels seem to be mediated by feelings of helplessness and uncertainty. Those subjects living near the dump site who reported the most helplessness and reported feeling the most uncertain about their situation also exhibited the most stress.

These results further support the notion that technological disasters may pose unique problems for their victims. In particular, the situation at TMI and the toxic dump site are similar because they both involve the possibility of exposure to a hazardous substance. In addition, data from both groups suggest that this type of exposure generates a great deal of uncertainty.

With toxic accidents becoming more common, it is important to study the psychological and physiological sequelae of this type of incident. The residents at TMI do not even know if they were exposed to radiation, yet they have continued to exhibit symptoms of stress for years. We can only imagine what the consequences would be if radiation or other toxic substances were found in their drinking water or in areas around their homes. Continuation of this type of research should prove valuable because it will provide additional information about technological and natural disasters and may provide insight on how to deal with the deleterious effects on residents who live near the sites of such accidents.

ACKNOWLEDGMENTS

This research was facilitated by support from the Uniformed Services University of the Health Sciences (CO7216) and the National Science Foundation (BNS8317997). The opinions or assertions contained herein are the private ones of the authors and are not to be construed as official or reflecting the views of the Department of Defense or the Uniformed Services University of the Health Sciences.

REFERENCES

Amkraut, A., & Solomon, G. F. (1977). From the symbolic stimulus to the pathophysiologic response: Immune mechanisms. In Z. J. Lipowski, D. R. Lipsitt, & P. C. Whybrow (Eds.), *Psychosomatic medicine: Current trends and clinical applications*. New York: Oxford University Press.

Baum, A., Fleming, R., & Davidson, L. (1983). Natural disaster and technological catastrophe. *Environment and Behavior, 15*(3), 333–354.

Baum, A., Fleming, R., & Singer, J. E. (1982). Stress at Three Mile Island. Applying psychological impact analysis. In L. Bickman (Ed.), *Applied social psychology annual* (Vol. 3, pp. 217–248). Beverly Hills, CA: Sage.

Baum, A., Fleming, R., & Singer, J. E. (1983). Coping with victimization by technological disaster. *Journal of Social Issues, 39*, 117–138.

Baum, A., Gatchel, R. J., & Schaeffer, M. A. (1983). Emotional, behavioral, and physiological effects of chronic stress at Three Mile Island. *Journal of Consulting and Clinical Psychology, 51*, 565–572.

Behavioral Medicine Special Report. (1979, May). Stress and nuclear crisis. *Behavioral Medicine*.

Bromet, E. (1980). *Three Mile Island: Mental health findings*. Pittsburgh: Western Psychiatric Institute and Clinic and the University of Pittsburgh.

Collins, D. L., Baum, A., & Singer, J. E. (1983). Coping with chronic stress at Three Mile Island: Psychological and biochemical evidence. *Health Psychology, 2,* 149–166.

Derogatis, L. R. (1977). *The SCL-90 Manual I: Scoring, administration, and procedures for the SCL-90.* Baltimore: Johns Hopkins University School of Medicine, Clinical Psychometrics Unit.

Dohrenwend, B. P., Dohrenwend, B. S., Kasl, S. V., & Warheit, G. J. (1979). *Report of the Task Group on Behavioral Effects to the President's Commission on the Accident at Three Mile Island.* Washington, DC.

Durrett, L. R., & Ziegler, M. G. (1980). A sensitive assay for catechol drugs. *Journal of Neuroscience Research, 5,* 587–598.

Fleming, I., & Baum, A. (1985–1986). The role of prevention in technological catastrophe. *Prevention in Human Services, 4,* 139–152.

Flynn, C. B. (1979). *Three Mile Island telephone survey.* Washington, DC: U. S. Nuclear Regulatory Commission (NUREG/CR-1093).

Flynn, C. B., & Chalmers, J. (1980). *The social and economic effects of the accident at Three Mile Island.* Washington, DC: U. S. Nuclear Regulatory Commission (NUREG/CR-1215).

Glass, D. C., & Singer, J. E. (1972). *Urban stress: Experiments on noise and social stressors.* New York: Academic Press.

Houts, P., Miller, R. W., Tokuhata, G. K., & Ham, K. S. (1980). *Health-related behavioral impact of the Three Mile Island nuclear accident.* Hershey, PA: Report submitted to the TMI Advisory Panel on Health-related Studies of the Pennsylvania Department of Health.

Jacobs, J. (1961). *The death and life of great American cities.* New York: Vintage Books.

Kraybill, D., Buckley, D., & Zmuda, R. (1979). *Demographic and attitudinal characteristics of TMI evacuees.* Paper presented at the Pennsylvania Sociological Society meetings.

Lazarus, R. S. (1966). *Psychological stress and the coping process.* New York: McGraw-Hill.

Lazarus, R. S., Opton, E. M., Jr., Nomikos, M., & Rankin, N. (1965). The principle of short-circuiting of threat: Further evidence. *Journal of Personality, 33,* 622–635.

Rodin, J., Solomon, S., & Metcalf, J. (1978). Role of control in mediating perceptions of density. *Journal of Personality and Social Psychology, 36,* 988–999.

Ross, R., & Glomset, J. A. (1976). The pathogenesis of atherosclerosis. *New England Journal of Medicine, 295,* 369–377, 420–425.

Selye, H. (1936). A syndrome produced by diverse nocuous agents. *Nature, 138,* 32.

Selye, H. (1956). *The stress of life.* New York: McGraw-Hill.

Staub, E., Tursky, B., & Schwartz, G. (1971). Self-control and predictability: The effects on reactions to aversive stimulation. *Journal of Personality and Social Psychology, 18,* 157–162.

Symington, T., Currie, A. R., Curran, R. S., & Davidson, J. N. (1955). The reaction of the adrenal cortex in conditions of stress. In *CIBA Foundation of colloquia on endocrinology, Vol. 8. The human adrenal cortex* (pp. 70–91). Boston: Little Brown.

Weiner, H. (1977). *Psychology and human disease.* New York: Elsevier.

Wortman, C. B., & Brehm, J. W. (1975). Responses to uncontrollable outcomes: An integration of reactance theory and the learned helplessness model. In L. Berkowitz (Ed.), *Advances in experimental social psychology* (Vol. 8, pp. 277–336). New York: Academic Press.

4

Psychopathological Consequences of Exposure to Toxins in the Water Supply

Margaret S. Gibbs
Fairleigh Dickinson University

The study reported here found, in brief, that a group of individuals exposed to toxins in their water supply showed impaired psychological functioning when tested several years after the event. Eighty-eight members of a litigation group from Legler, New Jersey showed high experience of stress, low sense of control of their environment as measured by the Rotter (1966) Internal–External Locus of Control scale, and high levels of psychopathology on the Minnesota Multiphasic Personality Inventory (MMPI) and Beck Depression Inventory (BDI) (Beck, Ward, Mendelson, Mock, & Erbaugh, 1961). Such results were predicted because of the many other studies showing a relationship between stress and psychopathology (e.g., Dohrenwend & Dohrenwend, 1974, 1981), and the findings of psychopathology in victims of disaster. Exposure to toxins can be considered a disaster in Barton's (1969) definition of disaster as a collective stress experience in which many members of a social system fail to receive expected conditions of life from the system.

If other studies find persons who are exposed to high stress or diaster to be psychologically impaired, what is the importance of the study of persons exposed to toxins? First, exposure to toxins is a social problem of increasing generality, and it is important to know what its consequences are for psychological functioning. Second, although a "natural" experiment has problems in demonstrating causality (Cook & Campbell, 1979), it has some advantages as well. Kasl (1983) effectively argues against the current tendency of stress researchers to rely on retrospective schedules of life events as a way of avoiding the perils of natural experiments. He points out the confounding of life event schedules with memory, age, lifestyle, mood, etc., and suggests that the complexity of the interrelationship of stress, lifestyle, and pathology indicates the advantages of a different strategy, the intensive study of single naturalistic events and their effects. One also can argue that a natural experiment has value in the study of the effects of learned helplessness. Learned helplessness cannot be well studied in a laboratory setting where it is not ethical to induce actual helplessness. Failure experiences are not for the most part experiences of helplessness. Subjects exposed to toxins unavoidably have already been exposed to stress and helplessness, and thus allow us a valuable opportunity to study these phenomena.

Third, the characteristics of toxin exposure are in some ways different from those of other disasters or stress experiences, and make toxin exposure an interesting topic for comparative purposes. Green, Grace, Lindy, Titchener, and Lindy (1983) report different kinds of consequences for different classes of victims at a nightclub fire, and they also note the differences between their findings and those obtained at the Buffalo Creek flood. They suggest that comparative work between types of disaster is necessary if we are to identify the important situational and person variables that produce and mediate psychopathology.

Exposure to toxins is an example of what Erikson (1976) has called a "man made" disaster, and what Baum, Fleming, and Davidson (1983) have described as a "technological" disaster. That is, exposure to toxins results from the failure of technology or the failure of humans to control it, and as such it has somewhat different implications from those of natural disasters like earthquakes. Technological disasters themselves, however, differ widely in their characteristics. Modern war, a plane crash, the flood at Buffalo Creek (caused by human error), a nightclub fire and exposure to toxins might all be considered technological disasters but they have different kinds of impact, even in the extent of their implications about living in a society built upon technology. Rather than regarding exposure to toxins simply as a technological disaster, let us look at what characterized it in the present case, in Legler, New Jersey.

CHARACTERISTICS OF THE DISASTER

The litigants who formed the population for this study lived in the area of Legler, New Jersey, surrounding a landfill that began operation in 1973. In the following years, rashes and other health problems led to a concern in residents that toxins might be being dumped and contaminating the water supply. The town dismissed residents' fears, but in November 1978, residents were informed that their water was in fact polluted with hazardous chemicals and should not be drunk or bathed in. For a period of a year, bottled water was hauled to residents. In 1979, hook-ups to city water began. The testing in the present study was carried out in March 1982.

Extent of Stressors

The description of stressors that follows comes from an interview study conducted by Edelstein (1982) on 25 of the Legler families. This study was available to me prior to my own research, and I am indebted to it.

The event lacked the widespread death and destruction associated with floods, fires, war, and so forth. However, a pervasive disruption of subjects' lives occurred, including the following.

Death, Illness and Concern About Illness.   This was the primary stressor. Although no deaths could be unequivocally proved to have been caused by the water, residents made this attribution for several deaths that occurred. Almost all respondents feared future health problems. A subject in this study conveys the sense of the fear in the following:

> I believe the toxic water crisis affected my life emotionally and psychologically in the following manner. I reason to believe as a result of drinking the water, it caused me to have a child born with defects. Which caused the child to pass away a few days after birth. This later caused problems between my wife and I, which lead to divorce. The effect of my first child death still fills me with a fear that my future child of my new wife could also end up in an unhappy experience. This child is due in the next 2 weeks. I also have to depress the fear that my life has a good opportunity of being shortened as a result of drinking the toxic water.

Financial Concerns.   Because property owners could not sell their homes for anything like what they had paid for them, if they moved out of the area, they stood to lose what was often their only equity or investment.

Using Bottled Water.   Hauling the water was difficult. Heating water for baths was difficult. Bathing children was a special problem. Some residents continued to shower in the polluted water because of the difficulties. The question of whether it was necessary to use the bottled water for such chores as washing the clothes, washing the dishes, scrubbing the floor the children would play on, and so forth was not easily answered and caused concerns.

Family Arguments.   The decision about whether or not to move, or when to use the bottled water, often led to arguments. If one parent wanted to move and the other didn't, and then sickness developed in a member of the family, the conflict could be grave.

Aesthetic Issues.   In addition to toxicity, the landfill posed other problems. In particular, odors developed, such that some subjects did not ever spend time outdoors or open their windows. Some residents stopped entertaining because of embarrassment about the odors and the difficulties with using bottled water.

Looking at the extent of the stressors, one would expect less severe emotional consequences than in a disaster with widespread death and grieving. However, one would expect to find a great deal of anxiety about health in the Legler subjects, as well as individualized responses to the stress that had many different kinds of impact for different individuals.

Long-Term Nature of the Stressors

Health concern would continue throughout the lives of the subjects and their children; the subjects knew that it could take 20 years for cancer to develop as a consequence of exposure. One would therefore expect long-term effects of the exposure; stress responses could appear even though testing took place several years after the period of greatest stress.

Lack of Definite Information About the Event

The events surrounding the exposure had no clear beginning, because almost certainly residents had been exposed prior to their receipt of notice. The toxins themselves were invisible; there was no way of knowing when one was exposed to them. Further, there was little clear-cut scientific information on how much one had ingested or what the later health hazards were likely to be. The lack of definite information was cited by several respondents as causing increased anxiety, and one would expect such to be the case. The lack of information also, however, makes the process of denial more possible in subjects.

Issues of Control

The lack of definite information available and the long-term nature of the consequences could be seen as adding to residents' sense of helplessness and lack of control. There was little or nothing they could do to attempt to forestall future health consequences for themselves or their children. Litigation could be perceived as a way of attempting to establish some control, but it was clear from subjects' remarks that they did not all find the litigation process satisfying. They found it demanding, protracted, and several mentioned that they did not feel money compensation was very relevant to their health concerns.

Baum et al. (1983) state that technological disasters involve a loss of expected technological control, in contrast to the lack of control experienced in a natural disaster such as a tornado. That is, technological disasters imply the failure of the technology that our society is built on, and one might expect such failure to result in a more generalized sense of helplessness than would occur in natural disaster. Janoff-Bulman and Frieze (1983) discuss the process of victimization, including exposure to toxins, as involving changes in the basic assumptions the individual has about the world. They isolate three such sets of beliefs: beliefs in one's invulnerability, belief that the world is meaningful, and a positive belief in the self. The victim must answer the fundamental question, "Why me?" and the answer involves a shift in these basic assumptions about one's self and the world.

One might well expect that exposure to toxins is the type of technological disaster that would require very fundamental shifts in one's perception of the world. That is, we know that lightning storms are dangerous and could kill us, or that technological failures such as plane crashes occur, but we do not expect drinking from the water faucet to be lethal. If drinking the water does prove lethal, a wider set of our assumptions about the world and our capacity to predict and control elements of it are violated than in natural disaster or many other forms of technological disaster. One would expect perceptions of the self as helpless, as not worth very much, and a perception of the world as uncontrollable and unpredictable. The study hypothesized that helplessness would be reflected in an external locus of control, and that in keeping with the learned helplessness model of depression, subjects would show high levels of depression.

That at least one subject felt such shifts in self-esteem and the meaningfulness of his life is shown in the following:

> I am a very disappointed man since we fined out our situation with the water situation, because since then many things happened in my life. I got so angry to kwik and many times I forget myself, and my wife and I live separately since Feb. 1980. I built my home with all my knolege and I did the very best to prove god and people, I did it myself from the footing to the decorating, which is including Masonry, Carpentry, Sheet-rocking, Flooring, Aluminum sideing, Painting and Paperhanging, Plumbing and Electric. I was so happy with my wife and I was a proud man. I lost this proud, and I allway thinking why I build that home for us, when after the TWP, was notified us, dont use our well water for further notice, because our well can be poisoned.

Social Support Within the Group

Social support is widely believed to have an ameliorating effect on reactions to stress (e.g., Dean & Lin, 1977). The presence of other victims during disaster can make the individual feel less personally victimized. A social network can provide both material and emotional support to the victim. The victim who has close friends or relatives with whom to share feelings and thoughts can obtain a cathartic release, practical advice about dealing with the associated problems, and opportunity to modify expectations and cognitive assumptions (Janoff-Bulman & Frieze, 1983) in a consensual fashion.

The Legler subjects did not for the most part feel they received social support from the community at large. A common response of subjects was to say that they felt betrayed or that they felt less trust in others as a consequence of their experience. Such feelings of anger and mistrust were directed toward those responsible for their experience, but appeared to have generalized. One would expect such loss of trust to be common in technological disasters, that is, disasters that can't be conceptualized merely as "acts of God." In fact, loss of trust seems common and important enough psychologically to warrant inclusion with the triad of shifted cognitive assumptions that Janoff-Bulman and Frieze (1983) discuss as a consequence of victimization. At any rate, different from disasters in which help comes pouring in from the outside, many Legler subjects felt alienated from the wider society. Two quotes from subjects in the present study follow:

> It has taught me not to trust people as much as before knowing the landfill had been responsible for the many times I have been sick. I hate them. I fear for my children, and the worry is always there.

> This incident has brought to bear many feelings questionable about the future. I feel that we have been needlessly victimized, yet it is far from unique in the times we live in. It has definately increased my sense of despair, personally as well as for others innocent as myself. Children particularly, as usual, have no choice but to fall prey to the callous disconcern of people whose decisions are made with only immediate gradification in mind. It is this warped sense of individualism that has identified itself with personal survival regardless of externalities that has caused this and other tragedies to happen.

Subjects also felt alienated from other individuals in the affected area who did not join the lawsuit. There was conflict within the community over this issue; for instance, some subjects reported that their children got into fights at school because they were labeled as part of the litigation. The litigation itself did form a bond for subjects that they reported helpful. Social support on the family level often was eroded by the arguments over selling out, and so on; family conflict was listed as one of the common sources of stress in the situation. In summary, there were no especially strong sources of social support to ameliorate the stress for Legler subjects.

Litigation as an Experimental Issue

Subjects were a special group in that not all exposed individuals chose to take legal action. Litigation groups have the advantage of allowing a researcher access to a group that might otherwise be unmotivated for testing; research on other litigation groups of disaster victims has provided useful findings (e.g., Green, Grace, Crespo da Silva, & Gleser, 1983). The possible danger to research, of course, is that subjects may be motivated to present themselves in a negative light to increase their chances of winning damages. In consideration of that possibility the following precautions were taken.

The study was conducted and described to subjects as a survey of the group as a whole, a survey in which no one individual's scores would have much impact on the average. Subjects were assured individual anonymity. A system of numbers was devised so that subjects could later identify their protocols should they desire feedback or to use the data in the litigation, but no one else would be able to identify their protocols.

Subjects were urged to answer honestly and told that the instruments used had scales that could detect when subjects presented themselves in an overly negative or positive fashion. The MMPI was chosen as the chief assessment instrument because of its scales for detecting individuals who "fake bad." The F minus K score was used to eliminate the MMPI scores of two subjects who did appear to be faking bad or answering randomly, whereas individuals with "fake good" profiles were retained in the data analysis.

MAJOR HYPOTHESES OF THE STUDY

Based on the aforementioned considerations, the following hypotheses were tested. It was predicted that subjects would describe their experience as stressful. An elevated external locus of control was predicted for subjects. High levels of health concern were predicted, as measured by the Hypochondriasis scale (Hs) of the MMPI; factor analysis of that scale has found almost all the variance to be accounted for by a single factor, health concern (Comrey, 1957b). Given the subjects' experience of helplessness, depression was predicted, as measured by scores on the Beck Depression Inventory (Beck et al., 1961) and the Depression (D) scale of the MMPI. Because different kinds of stressors were involved for different subjects, generalized psychopathology was predicted, as measured by the mean of the clinical scales of the MMPI.

METHOD OF THE STUDY

A letter from the law firm was sent to the 96 families in the suit at that time, requesting participation in a survey and offering feedback on the status of the suit. On the date specified, 88 persons, 48 males and 40 females, appeared and participated. The range of age of the sample was 20–78, with a mean of 38. The age distribution within the sample was equivalent to that within the adult members of the population, as was the length of residence in Legler for the sample and population.

Time constraints prohibited the construction of a control group; subjects'

scores were compared to norms for the measures used. The following measures
were administered:

Open-Ended Question.   Subjects were asked to describe in their own words
the impact, if any, of the toxic water crisis on their emotional well being and
that of their children. The degree of stress experienced by the individual or
family was rated on the 7-point scale presented as Axis IV of the Diagnostic
and Statistical Manual-III (American Psychiatric Association, 1980). This scale
allows for objective rating of events from the Holmes and Rahe scale (1967),
like death of a relative, illness of a relative, and so on. The open-ended questions
were rated by the experimenter and a research assistant; a reliability of .87 was
obtained.

Subjects' emotional responses to the stress were coded into categories. The
writer and the research assistant attained an agreement of at least 80% on all
categories except "worry" which tended to overlap with specific types of worry
and was therefore not included in the results.

Internal–External Locus of Control (I-E) (Rotter, 1966).

Beck Depression Inventory. (BDI) (Beck et al., 1961).

Minnesota Multiphasic Personality Inventory (MMPI) (Hathaway & McKinley,
1940).

Unreported Measures. Subjects wrote stories to two cards from the Thematic
Apperception Test shown on the slide projector. Findings from these stories were
to be used in the development of a new measure of ego functioning to be reported
at a future time. The Bender–Gestalt (Bender, 1938) was also administered in
a group form. Experimenter and research assistant could not attain adequate
reliability with the scoring system employed (Hain, 1964), and therefore hypotheses
and results regarding the Bender have been omitted.

The tests were administered to subjects at long tables in the local firehouse
by the experimenter and two assistants. The study was described as pertaining
to ways different individuals deal with stress. Points mentioned earlier regarding
confidentiality, research on a group versus on an individual, scales that detect
individuals who portray themselves too positively or negatively, and the importance of honesty were emphasized. The cover sheet repeated these points in
different terms.

Most subjects took the tests in the standard fashion. A small but vocal minority, however, demonstrated a good deal of hostility, as indicated by jokes and
complaints about psychologists, psychology, and the length and meaninglessness
of psychological tests. The complaints from subjects centered mainly on the fear
that pathology would be found when none existed, or that subjects' concerns
about the toxins would be attributed to being "crazy" rather than seen as realistic.
There were no jokes or comments about trying to look disturbed.

Complaints were dealt with good-naturedly, and subjects were told they did not have to complete the tests if they chose not to. Some such individuals did not finish particular tests. Other subjects who did not complete all of the tests complained of vision or comprehension problems, and others appeared to have difficulty with reading.

RESULTS

Demographic Variables

There were no significant correlations between either age or sex and the I–E, BDI, or MMPI mean of clinical scales. The highest correlation obtained was .18, $p < .10$, between the BDI and age, such that older subjects were more depressed. Similarly, there was no correlation between length of residence in Legler or distance from the landfill and the pathology or I–E measures. Because no significant correlations were obtained, these variables were not further considered in the analyses.

Stress

Table 4.1 shows the level of stress experienced by subjects on the DSM-III scale. Fourteen percent of the sample did not respond to this question, and an additional 5% responded in terms too general to rate. The mean level of stress obtained by this measure was 4.2 (i.e., a "moderate" level of stress with examples such as illness or family quarrels).

Table 4.2 shows the coding of the 76 subjects' answers about their emotional responses. Health worry was cited by almost half of the subjects; other common responses were a general mention of disturbance or upset, anger, depression, family quarrels, and mistrust of others.

I–E Scale

Of the 88 subjects, 82 completed the I–E scale. The mean score of the sample was 10.2, with a standard deviation of 4.9. Higher scores are more external. Table 4.3 shows the norms for the I–E scale for different groups cited by Lefcourt (1982), excluding those from foreign countries, pathological groups (i.e., prison or mental hospital inmates), groups chosen as especially internal or external (e.g., black students active vs. inactive in civil rights groups), and those groups in which a standard deviation is not provided. The Legler group's score is more external than all but two of the samples included. The Legler mean is significantly more external than the three most appropriate comparison samples included: Lichtenstein and Keutzer (1967), a sample of smokers more comparable in age

---

TABLE 4.1
Ratings of Stress Level from the Open-Ended Question

Stress level	% of 72 Ss[a]
1 no stress	4
2 minimal stress	1
3 mild stress	1
3.5[b]	4
4 moderate stress	44
4.5	19
5 severe stress	18
5.5	1
6 extreme stress	6

[a]Rounding accounts for total percentage equalling 98.
[b]Ratings are the averge of the two raters and thus may fall between two points on the scale.

TABLE 4.2
Subjects' Emotional Responses from the Open-Ended Question

Type of response	% of subjects
Worry over health	43
General disturbance, stress, or upset	34
Anger, temper outbursts, or hatred	24
Depression	17
Family quarrels	16
Mistrust of others	13
Financial worry	12
Feeling trapped or helpless	9
Divorce or separation	7
Nervous breakdown	4
No reaction	4

Note: N = 76.

TABLE 4.3
Scores from Normal Groups on the I-E Scale, from Lefcourt (1982)

Group	N	M	SD
Undergrads (Levy, 1967)	48	9.77	4.11
Smokers, M age 40.1 (Lichtenstein & Keutzer, 1967)	213	7.00	3.50
College males (Zytowski, 1967)	62	6.82	2.49
Undergrads in psychology (Feather, 1968)	46M	9.80	1.42
	88F	11.44	1.69
Undergrads (Hamsher, Geller, & Rotter, 1968)	60M	10.10	3.95
	113F	11.00	3.96
Undergrads, no sex differences (Julian & Katz, 1968)	1338	8.40	4.12
High school students (Hseih, Shybut, & Lotsof, 1969)	239	8.58	3.89
Female undergraduates (Crego, 1970)	99	7.97	3.80
Female student nurses (Lefcourt & Steffy, 1970)	37	7.14	3.28
Female undergrads (Strickland, 1970)	180	8.34	3.85
Administrators (Harvey, 1971): 1–10 years	21	7.19	2.75
11 years	27	5.41	3.15
Male undergrads (Lefcourt & Teleghi, 1971)	90	8.16	4.38
Males in Intro. Psychology (Phares, 1971)	646	9.20	3.48

Note: Normal groups included are those non-patient groups not selected for being likely to be internal or external (e.g., civil rights activists are omitted), and for whom standard deviations were provided. Higher scores are more external. References may be found in Lefcourt, 1982.

(M = 40.1) than the other samples [$t(293)$ = 6.25, $p < .001$]; Julian and Katz (1968), the largest sample of undergraduates [$t(1418)$ = 3.80, $p < .001$] and Phares (1971), the second largest sample of undergraduates [$t(726)$ = 2.33, $p < .05$].

BDI

Of the 88 subjects, 86 completed the BDI. The mean for the group was 11.54, with a standard deviation of 9.88.

Oliver and Simmons (1984) compared the BDI to DSM-III diagnoses of depression based on interviews for a randomly selected sample of 298 adults. They reported a cutting score of 10 correctly identified all individuals diagnosable as currently depressed, with a false positive rate of 13.7. Using a cutoff of ten, 51% of the Legler sample scored as depressed. Subtracting for false positives would still leave 44% of the group scoring as clinically depressed.

Comparing the Legler findings on the BDI to those obtained by Baum, Gatchel, and Schaeffer (1983) at Three Mile Island, we see that the Legler group was much more depressed than the TMI residents living within a 5-mile radius of the nuclear generator at the time of the radioactive leakage. The latter group scored a mean of 6.0 (SD = 6.5) on the BDI [$t(122)$ = 3.17, $p < .005$]. The Legler group also scored as significantly more depressed than three control groups used by Baum et al.; one living near an undamaged nuclear reactor, one near a coal-burning plant and one having no generating plant. These groups scored means on the BDI, respectively, of 3.50 [SD = 4.2, $t(116)$ = 4.45, $p < .001$]; 3.54 [SD = 3.6, $t(108)$ = 3.88, $p < .001$]; and 3.64 [SD = 3.3, $t(111)$ = 4.0, $p < .001$].

Other studies using the BDI on normal populations find scores in the range obtained by Baum et al. For instance, Bumberry, Oliver, and McClure (1978) found 16 college students judged normal in psychiatric interviews to show a mean of 3.94 (SD = 4.46) on the BDI, significantly lower than the Legler mean [$t(100)$ = 3.02, $p < .005$]. Gallagher, Nies, and Thompson (1982) found the mean for a sample of 82 normal persons over 60 years of age to be 5.54 (SD = 4.67) and 4.61 (SD = 4.84) in two administrations. Both of these scores are significantly lower than the Legler means [$t(166)$ = 5.0, $p < .001$ and $t(166)$ = 5.73, $p < .001$]. In a study of normal persons under stress, Nielsen and Williams (1980) found 8.9% of a sample of hospitalized medical patients scored over 15 on the BDI, in contrast to 28% of the Legler sample who scored above 15.

MMPI

Table 4.4 shows the means and standard deviations of the Legler subjects for the eight clinical scales of the MMPI. Of the 88 subjects, 76 completed all 399 items with no more than 30 omissions. An additional two subjects were excluded from the analysis on the basis of F minus K scores of over 20. One protocol

TABLE 4.4
Mean Scores on the MMPI Scales

Scale	M	SD
L	50.43	8.72
F	56.22	9.88
K	49.31	8.38
1 (Hs)	57.15	13.95
2 (D)	61.55	14.21
3 (Hy)	57.92	12.26
4 (Pd)	57.18	13.15
6 (Pa)	56.38	11.25
7 (Pt)	55.97	12.37
8 (Sc)	55.60	14.85
9 (Ma)	55.93	11.07
Mean of 1–4, 6–9	57.21	12.89

with an F minus K of 18 and one with an F minus K of 11 were retained because the general elevation of the profiles was not very high, nor did they have the configuration of faked profiles. No other protocol had an F minus K of more than 7.

The issue of comparison norms for the MMPI would appear to be more straightforward than for the other measures used, because the test was developed by differentiating the scores of normals and pathological groups, and then setting normal scores at a mean of 50 with a standard deviation of 10. The present study, however, precipitated scrutiny of the MMPI scores of normal groups, with the finding that normals tend to score somewhat higher than 50. Hsu and Gibbs (1986a) discovered, in studying the development of the MMPI, that there apparently was some overlap between the normal population used for the original item selection and the normal cross-validation population whose scores were used to develop norms. Such an error in test construction would mean that the cross-validation process would not correct for as much of the sampling error as it would have had independent samples been used for both pathological and normal groups, and that the selection of items to differentiate the two groups would not be optimum. Thus, the amount of separation of normal and pathological groups actually achieved by the test in its current use would be less than what the norms would predict; i.e., normal subjects should show scores slightly elevated over 50. This explanation would account for the slightly elevated scores normals in fact do show on the MMPI.

Table 4.5 provides a list of MMPI scores obtained by nonpsychiatric groups, excluding adolescents and the elderly (from Hsu & Gibbs, 1986a). It is apparent that normal groups generally score between 47.5 and 55, with the only exceptions being two small samples of Australian actors and Lutheran ministers. Normal groups under stress scored between 50 and 60. The mean of the clinical scales for the Legler subjects was 57.2. Because of the representativeness of the sample, the best scores to take for the purpose of normative comparison in the present study are those obtained by Colligan, Osborne, Swenson, and Offord (1983), that is a mean of 53.8 for a sample of 640 census-matched subjects. The Legler mean of the clinical scales is significantly elevated in comparison with this mean $[t(712) = 2.69, p < .01]$.

As predicted, scores on the Hypochondriasis (Hs) and Depression (D) scales were elevated, as compared to means obtained by Colligan et al. (1983). The Colligan et al. sample scored 52.5 $(SD = 10)$ on the Hs scale, as compared to the Legler score of 57.15 $(SD = 13.95)$ $[t(712) = 3.62, p < .001]$. The Colligan et al. sample scored 54 $(SD = 10)$ on the D scale, as compared to the Legler score of 61.55 $(SD = 14.21)$ $[t(712) = 5.85, p < .001]$.

As can be seen from Table 4.4, standard deviations of the MMPI clinical scales were higher than the normative 10 in the Legler sample. F-test comparisons indicate significantly higher standard deviations for all scales except Ma and Pa. $(F_{Hs} = 1.95, F_D = 2.02, F_{Hy} = 1.50, F_{Pd} = 1.73, F_{Pa} = 1.26, F_{Pt} = 1.52,$

TABLE 4.5
Mean MMPI Scores Obtained by Nonpsychiatric Groups

Source	Sample	Age	N male	N female	mean[a] male	mean[a] female	total
A. Normal Groups							
Leon, Gillum, Gillum, & Gouse, 1979	Physically and psychologically healthy men of upper middle or upper class	M47 (tested 1947)	71		48.07		
Spiaggia, 1950	Males of diverse occupations, not art students, mean IQ 112.7	M24.62	50		51.66		
Goodstein & Dahlstrom, 1956 as cited in Liverant, 1959	Parents of physically normal white children with no known behavioral problems. Volunteers from PTAs, etc.	males 40.8 females 37.3	49	49	53.25	52.25	
Anderson, 1969	Parents of Caucasian boys in fourth to sixth grade. Natural parents, married and living together.		50	50	52.75	53.75	
Colligan, Osborn, Swenson & Offord	Census-matched sample	M44	305	335	54.50	53.12	
B. Occupational Groups							
Schmidt, 1945	Normal soldiers, mean IQ 114.3	M24.9	98		47.46		
Vermiaud, 1946	Females in clerical, sales and optical occupations	med. 30–39	97		49.84		
Wiener, 1952	WWII veterans who received educational or vocational counseling. No data on combat experience.	M24.28	100		54.36		
Eschenbach & Dupree, 1959	Air Force officers and airmen on the day prior to a survival experience.	M29	22		54		
Kingsley, 1960	Enlisted men, on duty at a military installation, with no criminal history, matched with a criminal sample.		50		55.03		

Study	Description	Age				
Taft, 1961	Australian actors, average 14 years in theater, 59% unmarried.	half in 20s	46	28	59.69	57.12
Lucero & Currens, 1964	Lutheran ministers prior to 12-week clinical training period including psychotherapy. No information on selection procedure.		37		58	
Butcher, 1977, as cited in Butcher, 1979	Applicants for plant jobs.					51.12
King, Carroll, & Fuller, 1977	Full-time white male employees of a large company in the Midwest.		56		54.36	

C. Nonpsychiatric Samples under Stress

Study	Description	Age				
Wiener, 1952	WWII veterans disabled by asthma, arthritis, flat feet, gunshot wounds, heart disease, malaria, or skin disease	M mid to late 20s	450		55.82	
Linde & Patterson, 1958	Severely disabled cerebral palsy patients, half ambulatory, 3 with full use of hands	M24	14	19		56.8
Harrison & Kass, 1967	White lower-class pregnant women, patients at a public hospital.			389	56	
Swenson, Pearson & Osborne, 1973	Patients at Mayo Clinic. Some examinations were routine, but group shows elevated health concern on scales 1, 2, and 3.	med. 50–59	24277	25723	56.73	56.42
Sorbel & Worden, 1979	Hospitalized patients tested within 10 days of a Dx of cancer.	M50.2	56	74		
McGill, 1980	White female recipients of Aid to Dependent Children	M26		78	60.05	55.14

Note: Adolescent, aging, and black samples have been omitted. Studies are listed in chronological order. References are available from the author. "Mean of clinical scales 1 through 4 and 6 through 9.

$F_{Sc} = 2.20$, and $F_{Ma} = 1.22$. With $df = 73,697$, $F_{.05} = 1.32$). The high standard deviation is a reflection of the large number of highly elevated and depressed scores. Table 4.6 shows the number of subjects who scored at 70 or above on each of the clinical scales. For comparison purposes, we can use the data of Colligan, Osborne, and Offord (1980) who calculated the percentage of scores 70 or above obtained by the Hathaway and Briggs (1957) sample on the MMPI. (The exact composition of the Hathaway and Briggs, 1957, sample is unclear; it appears [Hsu & Gibbs, 1986a] to be the original Minnesota sample.) This data is provided in Table 4.4, and it is apparent that many more of the Legler subjects show elevations on all of the scales. On the Depression scale, 26% of the subjects had elevations of 70 or higher.

The proportion of subjects with at least one score at or above 70 was 48.6. Hsu and Gibbs (1986b) discuss the fact that the total proportion of subjects with scores at or above 70 will be affected by the intercorrelation of the scales, which will vary from sample to sample. For instance, imagine two samples, both of which show 20% of scores on each scale elevated to 70. If the first sample has scale intercorrelations of $+1.0$, then only 20% of this sample will show elevations of 70 or more, whereas if the second sample has low scale intercorrelations, then a much larger total percentage of subjects will show elevations of 70 or more. In general, the more highly intercorrelated the scales, the smaller the total number of persons who will show elevations of 70 or more, given an equivalent number of single scale elevations. Thus, data on scale intercorrelation should be provided when using this index, although it is seldom available (Hsu & Gibbs, 1986b). Table 4.7 provides the intercorrelation of MMPI scales with each other and the other measures. The mean intercorrelation of MMPI scales

TABLE 4.6
Percentage of Scores $\geq$ 70 on the Clinical MMPI Scales,
with Comparison Figures

Scale	% of Legler Scores $\geq$ 70	% Normative Scores > 70[a]
1 (Hs)	17.6	6.35
2 (D)	25.7	7.65
3 (Hy)	12.2	2.90
4 (Pd)	20.3	4.30
6 (Pa)	17.6	3.90
7 (Pt)	14.9	5.40
8 (Sc)	20.3	5.55
9 (Ma)	13.5	5.55

[a]Percentages are the means for the percentages of females and males with T scores > 70 provided by Colligan, Osborne, and Offord (1980). Colligan et al. state that using the criterion of scores $\geq$ 70 would raise the percentages in the table no more than 1 percentage point on three of the scales. Their sample is the Hathaway and Briggs (1957) group.

TABLE 4.7
Intercorrelations among MMPI Scales and Other Measures

	HS	D	Hy	Pd	Pa	Pt	Sc	Ma	MMMPI[a]	BDI	IE
D	.54**[b]										
Hy	.84**	.49**									
Fd	.48**	.57**	.56**								
Pa	.36**	.34**	.41**	.57**							
Pt	.62**	.69**	.56**	.69**	.56**						
Sc	.56**	.57**	.48**	.71**	.61**	.82**					
Ma	.29**	.22*	.28**	.45**	.43**	.56**	.54**				
MMMPI	.77**	.73**	.76**	.82**	.68**	.90**	.87**	.60**			
BDI	.46**[c]	.63**	.36**	.55**	.55**	.60**	.61**	.29**	.68**		
IE	.17[d]	.29**	.12	.21*	.10	.24*	.28**	.05	.25*	.35**[e]	
stress	.33**[f]	.16	.26*	.14	.15	.22*	.28*	.16	.28*	.33***[g]	.12[h]

[a]M MMPI = M of scales 1–4 and 6–9. [b]n = 74 for correlations between MMPI scales. [c]n = 72 for correlations between the BDI and MMPI scales. [d]n = 70 for correlations between IE and MMPI scales. [e]n = 82 for correlation between the BDI and IE. [f]n = 60 for correlations between stress level and MMPI scales. [g]n = 72 for correlation between stress level and the BDI. [h]n = 68 for correlation between stress level and IE.
*p < .05. **p < .01.

was .53. This is a higher value than for the normal samples reported in Dahlstrom, Welsh, and Dahlstrom (1975). Thus the high percentage of subjects in the Legler sample with scores at or above 70 cannot be viewed as an artifact of low scale intercorrelation.

Another index of severe pathology is the Goldberg Index. The mean for the Legler group was 48.5 (SD = 16). Scores over 45 have been used to discriminate psychotic profiles, or profiles of serious psychopathology (Graham, 1977). In addition, an examination of the individual records showed that 27% showed at least one of the following signs on the MMPI that are usually viewed as indicating a clinical diagnosis of mental disorder: D scale 80 or above; Hs and Hy 70 or above; Hs, D, and Hy 70 or above; Pa 75 or above; or Sc 80 or above.

Considering the number of elevated profiles, there were a number of low scores as well. Table 4.8 indicates the number of scores 40 or less on the MMPI scales. Little data is available on the number of low scores ordinarily occurring, although we do know that they are more unusual than high scores because of the "basement effect" in the test.

Intercorrelation of Measures

Table 4.7 shows the intercorrelation of measures with each other. The BDI and the D scale of the MMPI correlated .63. The BDI showed sizable correlations with most of the other MMPI scales and correlated .68 with the mean of the clinical scales.

TABLE 4.8
Percentage of Legler Subjects with
MMPI Scores ≤ 40

Scale	% ≤ 40
1 (Hs)	5.4
2 (D)	1.4
3 (Hy)	8.1
4 (Pd)	6.8
6 (Pa)	4.0
7 (Pt)	6.8
8 (Sc)	12.2
9 (Ma)	9.5

Stress ratings correlated significantly with the following pathology measures: .33 with the Hs scale ($p < .005$), .33 with the BDI ($p < .005$) and .28 with the mean of the MMPI clinical scales ($p < .05$). Stress ratings also correlated significantly with three of the other seven MMPI scales.

The I–E scale showed moderate correlations with the measures of depression: it correlated .35 with the BDI ($p < .001$) and .29 with the D scale ($p < .01$). Thus persons with highly external scores were more likely also to be depressed. The I–E scale correlated .25 with the mean of the MMPI clinical scales ($p < .005$) and with three of the other seven MMPI scales. The item clusters described by Reid and Ware (1973) for the I–E scale (Fatalism and Social System Control) both showed lower correlations than the total I–E scale with the pathology measures and are not reported.

DISCUSSION OF THE RESULTS

All the main hypotheses of the study were supported. Subjects reported their experience with the toxins as stressful. They showed elevated externality on the I–E scale. They showed high levels of health concern and a level of generalized psychopathology higher than the norm and consonant with the level reported by other groups undergoing environmental stress. They showed high levels of depression, both on the BDI and on the MMPI.

Stress correlated with the pathology measures at about the level reported in other studies (e.g., Pearlin, Menaghan, Lieberman, & Mullan, 1981). The "moderate" level of stress reported is probably an underestimate of the true amount of stress attributable to the experience. The question from which ratings were made asked subjects for their reaction to the water situation, rather than for their experience; subjects varied in how much detail they provided in recounting the stressors themselves. It should also be noted that the reported stress is that attributable to the toxic water experience itself, and not the total amount of life stress experienced by subjects.

The pathology reported by subjects included many other reactions besides depression and elevated health concern. For instance, looking at the percentage of scores 70 or above on the Pd scale of the MMPI (Table 4.4), 20% of scores were elevated, implying generalized anger toward authority figures and the possibility of acting out in these subjects. On the Pa scale, 18% of scores were elevated, implying mistrust and oversensitivity toward others, the possibility of external attributions as a way of denying and coping with self-blame, as well as some serious personality disintegration in the 8% of subjects who scored over 75 on this scale. On the Sc scale, 20% of scores were elevated to at least 70, implying high levels of generalized anxiety and distress, feelings of not being in control of aspects of the environment and the self, and some serious personality disintegration in some of the subjects (9% of subjects scored over 80 on this scale). Other scales also showed elevations higher than would be expected. The Legler group clearly showed a wide range of types of psychopathology.

It also is clear that some members of the Legler group showed serious pathology. To state simply that this group exposed to toxins showed a mean pathology level on the MMPI and BDI higher than the norm is to miss the importance of the number of exposed individuals who showed indications of serious pathology. Using Oliver and Simmons' (1984) criteria, 44% of the group appear diagnosable as clinically depressed on the basis of the BDI. On the MMPI, 48% of the sample showed at least one score at or above 70.

The psychopathology in the Legler group was clearly much greater than that found at that much-studied technological disaster, TMI. Legler BDI scores were significantly more elevated than those obtained by Baum et al. (1983) at TMI. In addition, evidence of clinical psychopathology at diagnosable levels was obtained at TMI primarily in studies conducted immediately after the event (e.g., Dohrenwend et al., 1981).

How can we understand the experiences of these individuals who do not show high levels of psychopathology? The intercorrelation of measures indicates that those individuals who experienced lower levels of stress and were more internal in their locus of control were less likely to show pathology. The size of these correlations, however, is quite small, indicating that stress level and locus of control have low levels of explanatory power in the Legler situation. One possible clue to the difference between the high and low pathology individuals is to note the high variability obtained in the pathology measures; subjects tended either to score quite high or quite low. As mentioned, 35% of the sample showed at least one score of 40 or less on the MMPI. Low scores on the MMPI are more difficult to interpret than high scores, but one possible explanation is that they indicate a pattern of defensiveness, conventionality, and denial. That is, some of these low scoring subjects may simply be unusually healthy, but others may be attempting to deal with the stress of their situation through denying its emotional impact. As mentioned earlier, the very indefiniteness of toxic water exposure as a stress experience may foster patterns of denial in victims. Collins, Baum, and Singer (1983) found that at TMI the use of denial as a strategy was less effective in reducing the consequences of stress than other strategies. Thus, there is some question as to whether the Legler subjects who score low on pathology are really functioning on a much higher level than those scoring high on pathology.

The acrimony in the community as to whether to become part of the litigation suggests that non-litigants also may have used denial as a strategy more than litigants. It also is possible that because all subjects were litigants, their experience may have been different than non-litigants who were exposed to the same toxic water situation. Although "faking bad" did not appear to be a problem in the group, judging from the test administration and the validity scales of the MMPI, it may be that litigants were more affected by the experience than non-litigants. That is, the level of pathology in non-litigants might be lower, with fewer highly elevated scores and more patterns of health and/or denial.

Another issue in the design of this experiment is the use of group norms for the tests for comparison purposes. One of the problems with the natural experiment is that the design must be quasi-experimental rather than experimental (Cook & Campbell, 1979). That is, one can never assign subjects randomly to be exposed to toxins or to a control group, but because random assignment is not possible, causality is more difficult to demonstrate. Test norms and other test scores for normal and comparative groups were used in this study to provide the control group. Although such a strategy may appear less satisfactory than using a matched control group, in fact matching poses many hazards of its own when randomization cannot be implemented (Cook & Campbell, 1979). The multitude of studies using the test instruments employed in this study makes for greater confidence that the Legler subjects' scores were indeed elevated above the norm.

Some of the present findings are relevant to the learned helplessness model of depression. Findings can be seen as offering support, because subjects showed a highly external locus of control reflecting the lack of control they experienced in the water crisis. Locus of control scores correlated with the depression measures, which also were highly elevated.

An alternative explanation of the findings is congruent to the model proposed some years ago by Kelly (1955). Kelly felt that the ability to predict and control one's environment is a basic human need. When one cannot predict or control, one becomes anxious, the extent of the anxiety depending on how central the areas of lack of control are to one's world view. (As mentioned earlier, an experience in which drinking water from the faucet turns out to be potentially lethal involves loss of control over aspects of experience very central to one's perception of the world.) The way individuals deal with their anxiety depends on their abilities, established patterns of defenses, and their construct system. Individuals with better problem-solving skills will in general find better ways of handling the unpredictability, although in this case there were no very good solutions to the problem of toxic exposure. Many kinds of emotional problems could result, including depression, generalized anxiety syndromes, patterns of anger and mistrust, and patterns of denial and exaggerated health. In contrast to the learned helplessness model of depression, the Kelly model assumes that the capacity to predict and control is central to all aspects of pathology and health. It also suggests that in developing a cognitive theory of psychopathology, we may need to focus on the cognitions of the individual, rather than trying to build a model of attributions held by all members of a diagnostic group.

As mentioned, many different patterns of psychopathology emerged in the present sample. The I–E scale was not only correlated with measures of depression, it also correlated significantly with the Pd, Pt, and Sc scales of the MMPI and with the mean of the MMPI clinical scales. Thus, helplessness and externality were linked to a more generalized pathology than depression alone. Of course, many other researchers have found low but positive correlations between externality and psychopathology (Lefcourt, 1982).

One can even question whether depression as a diagnostic entity, separate from other aspects of psychopathology, has heuristic value. The clinician knows how varied are the pictures of "depression" seen. In addition, there is a lack of discriminant validity for concepts like depression and anxiety. Watson and Clark (1984) have pointed out the high intercorrelations found between many common measures of pathology, including MMPI scales, the BDI and other anxiety and depression measures. They suggest that we should begin to think of "negative affect" as a construct, rather than our more traditional concepts. Similarly, Gotlib (1984) reports high intercorrelations between the BDI and 16 other scales of various forms of psychopathology in a large sample of college students. The present study found higher correlations between the BDI and non-depressive scales of the MMPI than with the Depression scale. Of course, the Depression scale of the MMPI itself is no more a "pure" measure of depression than the BDI; Comrey (1957a) described the first factor of the D scale as "Neuroticism."

If there is some question as to whether depression forms a clearly definable syndrome, surely it is premature to attempt to define cognitions held by all depressed individuals. This seems particularly true within the reformulation of the learned helplessness model (Abramson, Seligman, & Teasdale, 1978), with its statement that the individual must attribute the experience of lack of control to the self in order for depression to result. This involves a basic paradox; the individual is saying, "Things are out of my control, but it is my fault, that is, in my control." It is not surprising that the results are inconsistent as to whether depressed persons make internal or external attributions (Coyne & Gotlib, 1983). Both Wortman and Dintzer (1978) and Janoff-Bulman and Frieze (1983) state that self-attributions for an experience of victimization can result in lowered levels of depression. If you are making an internal attribution you are not really helpless. This seems to be the point that is missed in the reformulation model, that when you are truly helpless there are no attributions to make. In Kelly's theory, the event lies outside your capacity to conceptualize and understand it.

For instance, look back at the two quotes from individuals who describe the mistrust of others that developed as a consequence of their exposure to toxins. One is struck by the difference between the writers. The second writer has a highly developed theoretical explanation for what happened, whereas the first writer can only say, "I hate them." Although both writers seem much affected by their experience, one would expect the second writer to be less anxious and to show less pathology than the first because the experience is more understandable to this person.

Another example serves to show the importance of looking at the cognitions of the individual in understanding the stress response. The following quotation is an interesting one in showing a woman whose locus of control could be said to have become more internal through the stress of her experience.

> My children have lived in a constant state of upheaval since this crisis broke. There have been late dinners, phones ringing constantly, arguments, people coming and going at all hours. They have heard people discussing their past and future health problems. My son has expressed nightmares and fears about getting ill and dying like his sister has. His school teachers have told me that an argument at home over the water results in his causing troubles in class the next day. For me, I feel my whole life has changed. I have gone from a trusting citizen to one who fully doubts the government knows what it is doing. I have changed from a passive person to one who speaks my mind openly. I have gone from being a rosey eyed idealist to a full scale doubting thomas. I believe my future and my children's future *must* be overseen by me and I *must* make every effort to control the things that happen to us. I have come to believe the old adage that "God helps those who help themselves."

Whether the subject quoted changed in her level of adjustment is not possible to know, but the quote suggests that she may be functioning behaviorally on a higher level, at the cost of greater internal pain. At any rate, once one begins to look at the individual level of responses to stress and helplessness, it becomes apparent how different and complex are the mechanisms and cognitions involved. We need to develop better techniques and measures for examining these differences.

It is not surprising that the variables of stress and externality do not account for more of the variance in psychopathology, either in this study or others. From the perspective of experimental design, the present study has the weaknesses of other natural experiments. A natural experiment like the present one, however, can show us a broader, more complex and more real picture of the psychological response to stress than can a more restricted laboratory study.

REFERENCES

Abramson, L. Y., Seligman, M. E. P., & Teasdale, J. D. (1978). Learned helplessness in humans: Critique and reformulation. *Journal of Abnormal Psychology, 87,* 49–74.

American Psychiatric Association. (1980). *Diagnostic and statistical manual of mental disorders* (3rd ed.). Washington, DC: Author.

Barton, A. (1969). *Communities in disaster.* Garden City, NY: Doubleday.

Baum, A., Fleming, R., & Davidson, L. M. (1983). Natural disaster and technological catastrophe. *Environment and Behavior, 15*(3), 333–354.

Baum, A., Gatchel, R. J., & Schaeffer, M. A. (1983). Emotional, behavioral, and physiological effects of chronic stress at Three Mile Island. *Journal of Consulting and Clinical Psychology, 51*(4), 565–572.

Beck, A. T., Ward, C. H., Mendelson, M., Mock, J., & Erbaugh, J. (1961). An inventory for measuring depression. *Archives of General Psychiatry, 4,* 561–571.

Bender, L. (1938). A visual motor Gestalt test and its clinical use. *American Orthopsychiatric Association, Research Monographs,* No. 3.

Bumberry, W., Oliver, J. M., & McClure, J. N. (1978). Validation of the Beck Depression Inventory in a university population using psychiatric estimate as a criterion. *Journal of Consulting and Clinical Psychology, 46,* 150–155.

Colligan, R. C., Osborne, D., & Offord, K. P. (1980). Linear transformation and the interpretation of MMPI T scores. *Journal of Clinical Psychology, 36,* 162–165.

Colligan, R. C., Osborne, D., Swenson, W. M., & Offord, K. P. (1983). *The MMPI: A Contemporary normative study.* New York: Praeger.

Collins, D. L., Baum, A., & Singer, J. E. (1983). Coping with chronic stress at Three Mile Island: Psychological and biochemical evidence. *Health Psychology, 2,* 149–166.

Comrey, A. L. (1957a). A factor analysis of items on the MMPI depression scale. *Educational and Psychological Measurement, 17,* 578–585.

Comrey, A. L. (1957b). A factor analysis of items on the MMPI hypochondriasis scale. *Educational and Psychological Measurement, 17,* 568–585.

Cook, T. D., & Campbell, D. T. (1979). *Quasi-experimentation: Design and analysis issues for field settings.* Boston: Houghton Mifflin.

Coyne, J. C., & Gotlib, I. H. (1983). The role of cognition in depression: A critical appraisal. *Psychological Bulletin, 94,* 472–505.

Dahlstrom, W. G., Welsh, G. S., & Dahlstrom, L. E. (1975). *An MMPI handbook* (Vols. 1 & 2). Minneapolis: University of Minnesota Press.

Dean, A., & Lin, N. (1977). The stress-buffering role of social support: Problems and prospects for systematic investigation. *Journal of Nervous and Mental Disorder, 165,* 7–15.

Dohrenwend, B. S., & Dohrenwend, B. P. (Eds.). (1974). *Stressful life events: Their nature and effects.* New York: Wiley.

Dohrenwend, B. S., & Dohrenwend, B. P. (Eds.). (1981). *Stressful life events and their contexts.* New York: Prodist.

Dohrenwend, B. P., Dohrenwend, B. S., Warheit, G. J., Bartlett, G. S., Goldsteen, R. L., Goldsteen, K., & Martin, J. L. (1981). Stress in the community: A report to the President's Commission on the Accident at Three Mile Island. *Annals of the New York Academy of Science,* 159–174.

Edelstein, M. (1982). *The social and psychological impacts of groundwater contamination in the Legler section of Jackson, N.J.* Unpublished manuscript.

Erikson, K. (1976). *Everything in its path.* New York: Simon & Schuster.

Gallagher, D., Nies, G., & Thompson, L. (1982). Reliability of the Beck Depression Inventory with older adults. *Journal of Consulting and Clinical Psychology, 50,* 152–153.

Gotlib, I. H. (1984). Depression and general psychopathology in university students. *Journal of Abnormal Psychology, 93,* 19–30.

Graham, J. R. (1977). *The MMPI: A practical guide.* New York: Oxford University Press.

Green, B. L., Grace, M. C., Crespo da Silva, L., & Gleser, G. C. (1983). Use of the psychiatric evaluation form to quantify children's interview data. *Journal of Consulting and Clinical Psychology, 51,* 353–359.

Green, B. L., Grace, M. C., Lindy, J. D., Titchener, J. L., & Lindy, J. G. (1983). Levels of functional impairment following a civilian disaster: The Beverly Hills supper club fire. *Journal of Consulting and Clinical Psychology, 51,* 573–580.

Hain, J. D. (1964). The Bender Gestalt Test: A scoring method for identifying brain damage. *Journal of Consulting Psychology, 28,* 34–40.

Hathaway, S. R., & Briggs, P. F. (1957). Some normative data on new MMPI scales. *Journal of Clinical Psychology, 13,* 364–368.

Hathaway, S. R., & McKinley, J. C. (1940). A multiphasic personality schedule (Minnesota): 1 Construction of the schedule. *Journal of Psychology, 10,* 249–254.

Holmes, T. H., & Rahe, R. H. (1967). The Social Readjustment Rating Scale. *Journal of Psychosomatic Research, 11,* 213–218.

Hsu, L., & Gibbs, M. (1986a). *Bias of MMPI norms.* Manuscript submitted for publication.

Hsu, L., & Gibbs, M. (1986b). *Effects of differences in intra-group relations of MMPI scales on the proportions of primed MMPI profiles in groups.* Manuscript submitted for publication.

Janoff-Bulman, R., & Frieze, I. H. (1983). A theoretical perspective for understanding reactions to victimization. *Journal of Social Issues, 39*(2), 1–17.

Julian, J. W., and Katz, S. B. (1968). Internal versus external control and the value of reinforcement. *Journal of Personality and Social Psychology, 76,* 43–48.

Kasl, S. V. (1983). Pursuing the link between stressful life experiences and disease: A time for reappraisal. In C. L. Cooper (Ed.), *Stress research: Issues for the eighties* (pp. 79–102). Chicester, England: John Wiley.

Kelly, G. A. (1955). *The psychology of personal constructs.* New York: Norton.

Lefcourt, H. M. (1982). *Locus of control: Current trends in theory and research* (2nd ed.). Hillsdale, NJ: Lawrence Erlbaum Associates.

Lichtenstein, E., & Keutzer, C. S. (1967). Further normative and correlational data on the internal–external (I.E.) control of reinforcement scale. *Psychological Reports, 21,* 1014–1016.

Nielsen, A. C., & Williams, T. A. (1980). Depression in ambulatory medical patients: Prevalence by self-report questionnaire and recognition by non-psychiatric physicians. *Archives of General Psychiatry, 37,* 999–1004.

Oliver, J. M., & Simmons, M. E. (1984). Depression as measured by the DSM-III and the Beck Depression Inventory in an unselected adult population. *Journal of Consulting and Clinical Psychology, 52,* 892–898.

Pearlin, L. I., Menaghan, E. G., Lieberman, M. A., & Mullan, J. T. (1981). The stress process. *Journal of Health and Social Behavior, 22,* 337–356.

Phares, E. J. (1971). Internal-external control and the reduction of reinforcement value after failure. *Journal of Consulting and Clinical Psychology, 37,* 386–390.

Reid, D., & Ware, E. E. (1973). Multidimensionality of internal-external control: Implications for past and future research. *Canadian Journal of Behavioral Science, 5,* 264–271.

Rotter, J. B. (1966). Generalized expectancies for internal versus external control of reinforcement. *Psychological Monographs, 80,* (Whole No. 609).

Watson, D., & Clark, L. A. (1984). Negative affectivity: The disposition to experience aversive emotional states. *Psychological Bulletin, 96,* 465–490.

Wortman, C., & Dintzer, L. (1978). Is an attributional analysis of the learned helplessness phenomenon viable? A critique of the Abramson-Seligman-Teasdale reformulation. *Journal of Abnormal Psychology, 87,* 75–90.

5
Assessing the Impact of Hazardous Waste Facilities: Psychology, Politics, and Environmental Impact Statements

Kenneth M. Bachrach*
Alex J. Zautra
Arizona State University

Living near a hazardous waste disposal facility is a prospect few people readily welcome. Yet hazardous wastes such as radioactives, flammables, asbestos, and acids are being manufactured at a rate that far exceeds the capacity of existing storage and disposal facilities. It has been estimated that roughly 350 pounds of hazardous waste is produced annually for every American citizen (Epstein, Brown, & Pope, 1982). Compounding the problem is the disturbing realization that hazardous waste has been disposed of at thousands of illegal dump sites throughout the nation. Congress appropriated a $1.6 billion "Superfund" to clean up these illegal and dangerous dump sites, but these clean-up activites are preceding at a very slow rate.

New hazardous waste disposal facilities will need to be built in the coming years to accommodate this myriad of hazardous substances. Some of these new facilities will require transportation of hazardous substances along the public highway and rail systems. The media has sensitized the public to the problems associated with hazardous and toxic substances through the reporting of events at Love Canal and Three Mile Island (TMI). It is becoming increasingly clear that people do not want these potentially health-threatening substances located near them. Public opposition to hazardous waste facilities is well-documented (U.S. EPA, 1979).

Hazardous waste disposal facilities are perceived in a manner similar to penal institutions: There is a recognized need for such facilities, but nobody wants to live near them. Debate on where to locate these facilities classically illustrates

---

*Now at the School of Public Health, University of California, Los Angeles.

the principle of negative reinforcement. The hazardous waste facility is an aversive stimulus. People try to keep hazardous waste facilities as far away from them as possible, in order to terminate or lessen the intensity of this stimulus. Because the reinforcing aspects of these actions come from keeping the aversive stimulus at a distance, debate revolves around where not to locate the facilities. The hazardous waste facility is a "hot potato," which one wants to be rid of as soon as possible.

Yet what happens to the one party or community that is left holding the hot potato? What coping mechanisms do residents in these communities employ? Do they move away? Do they engage in social action? Do they display symptoms and behaviors characteristic of stress, anxiety, and depression? And what is the long-term psychological impact of living near a hazardous waste facility?

The answer to these questions requires knowledge of community residents before the hazardous waste facility is built. Yet most psychological studies of environmental hazards have been reactive in nature. They have occurred following a major calamity, such as after the accident to the nuclear reactor at Three Mile Island. By necessity they are often hastily designed in order to meet pressing time constraints. Some investigations (Bromet, Parkinson, Schulberg, Dunn, & Gondek, 1980; Dohrenwend et al., 1981) have employed matched control groups to assess whether residents living or working near the problem area differ on selected measures from those living or working further away. But the equivalence of these groups is impossible to guarantee.

Adequate assessment of the impact of a major change in the environment requires some knowledge of what the people who live in that environment were like before the change occurred. Dohrenwend et al. (1981) concluded that over 10% of the local population was adversely affected as a direct result of the TMI nuclear accident, because severe distress levels decreased from 26% immediately following the accident to less than 15% in the months to follow. An implicit assumption was that distress levels returned to proportions comparable to those before the accident. However, it is conceivable that only 5 or 10% of the population was severely distressed before the accident and that the 15% represents an elevated baseline level. On the other hand, another less likely scenario also is possible. People may have been strengthened by the manner in which they handled the crisis, so that the 15% baseline following the accident indicates a drop in distress from a pre-accident level of 20%. The point is that assessments that only take place following the environmental change (e.g., the nuclear accident) are flawed from an experimental design standpoint. The true "pre-levels" are never known, and thus a valuable piece of the puzzle is missing.

There are advantages in assessing the psychological impact of a planned hazardous waste facility on community residents. The fears and concerns of local residents can be identified, and it can be ascertained whether community leaders and vocal residents represent the community at large. Such assessments also begin a dialogue between state and local leaders and can prove productive if handled effectively. Once concerns have been identified, interventions can be developed to address these issues and ward off potential or emerging problems. Most important may be the psychological record such an assessment provides to serve as baseline data for future studies. Should real or perceived problems develop at a later point in time, concrete data exist for comparison purposes.

A logical point to collect such data would be during the environmental impact statement (EIS) study stage. Such studies could be funded through independent research grants or by the state or federal government as part of the EIS. Environmental impact statements are reports that assess and analyze the impact of proposed federal actions, and are used by regulatory agencies to decide whether the proposed actions should be allowed. They are an outgrowth of the National Environmental Policy Act (NEPA) of 1969. The aim of NEPA is to protect the environment as well as the health and safety of the general population in any federal environmental use project. The language of NEPA is vague in terms of what constitutes the health of the people this legislation is mandated to protect and promote. There is no exclusion of psychological health factors, yet psychologists are rarely consulted during the EIS process, and few psychologists have loudly voiced their desire to be included in the process.

Environmental impact statements usually measure aspects of the physical environment, but rarely do they examine the psychological impact proposed changes may have on people who live in the environment. There are a number of possible reasons for this lack of attention to and interest in psychological issues. First, many projects that require an environmental impact statement do not have a profound effect on people. Certain wildlife may be threatened along with artifacts of historical and cultural value, but the general population is largely unaffected. Second, persons involved in the EIS process often are not psychologically minded. People drafting the EIS may be engineers, chemists, and geologists. Health concerns are thought of in terms of physical health without an understanding of the debilitating effects of severe psychological distress. Third, the thorough assessment of psychological issues in an EIS is unprecedented. This seems due to the skepticism and legitimate concern among government officials of whether such an assessment can be adequately conducted, for a reasonable cost. Also, it is unclear as to how one would be conducted even if the interest was present.

The issue of considering psychological factors in an EIS came to the forefront after the nuclear accident at Three Mile Island. A group called People Against Nuclear Energy (PANE) argued to the Nuclear Regulatory Commission (NRC) that the restart of the TMI Unit 1 reactor would cause severe psychological distress to individuals living near the facility and be damaging to the well-being of the surrounding community. The NRC considered psychological factors in reviewing this case, but after a series of legal actions, the Supreme Court ruled (Metropolitan Edison Co. v. PANE, 1983) that psychological distress produced by perceptions of risk are not sufficient grounds to stop a federal environmental use project. The Court decided that psychological injuries can be considered under NEPA, but only when they are due to changes in the physical environment (see Hartsough & Savitsky, 1984, for a review of the TMI case).

The Court raised important questions that psychologists need to address. However, it is important that researchers' lines of inquiry into the psychological impact of environmental hazards not be solely determined by legal decisions. Whether or not psychological factors are legally mandated for inclusion (or exclusion) in an EIS is secondary to whether or not the psychological impact of a change in the environment can be accurately and reliably assessed.

This chapter aims to deal with these issues by presenting the methodology and findings of such a psychological assessment. However, before a description and discussion of the actual study can begin, some background information is necessary. Any researcher undertaking such an investigation must recognize that he or she is entering a well-greased system whose wheels have been turning for many years. By the time the decision to build a hazardous waste facility in a specific location has been made, many preliminary political and legal battles have been fought.

SITE SELECTION

In the late 1970s the Arizona Department of Health Services (ADHS) recognized the need to develop a hazardous waste disposal site. It was estimated that in 1980, 6.4 million gallons of liquid hazardous waste and 113,000 tons of solid hazardous waste were generated in Arizona, and that these figures were likely to increase 5–10% annually (ADHS, 1981). Much of the waste being generated within the state was not being disposed of by environmentally acceptable methods. ADHS claimed the most important reason for this was the lack of any hazardous waste disposal site or treatment facility in Arizona (ADHS, 1981).

ADHS brought this issue to the Arizona State Legislature, which authorized ADHS to begin the process of selecting a site for a hazardous waste facility (HWF) in August 1980. An interdisciplinary task force was established by ADHS to evaluate and select suitable sites. An initial screening produced 23 possible sites, with a more refined screening reducing the number to 10. These 10 sites were evaluated in depth and in January 1981 a report to the Arizona State Legislature was published recommending one site and deeming two others worthy of "strong consideration" (ADHS, 1981).

The Arizona State Legislature chose one of the two alternative sites, Rainbow Valley, as the location for the proposed HWF in February 1981. Legislative politics played a major role in the decision-making process (Kolbe, 1981). The most highly ranked site was in the corner of the district of the Senate Majority Leader, who had earlier accommodated a new state prison. The other alternative site was located in the district of the Senate Minority Leader, whose support often was needed, because the Republican majority controlled the Senate by a small 16 to 14 margin. The Democratic State Senator from Rainbow Valley had the least political clout of the three legislators.

Residents living in or near Rainbow Valley were very upset over the legislature's decision. They had lobbied long and hard against the selection of the Rainbow Valley site by attending meetings, writing letters, and circulating petitions. They were angry that the top recommendation, which was geographically more isolated and not close to any sizeable community, had not been chosen. A number of residents were quite cynical about the entire decision-making process, and viewed it as tied to an earlier decision to build a needed bridge connecting Rainbow Valley with a major highway. The local community had fought for many years for such a bridge, because in the winter a normally dry river bed would often swell with water, washing out the major transportation linkage for the community with the Phoenix metropolitan area. Local residents came to believe that a very sturdy and expensive bridge was finally built to accommodate not just local traffic but also large trucks that would carry hazardous waste materials.

Initial Planning for the Hazardous Waste Facility

After selecting the Rainbow Valley site, the Arizona Legislature authorized ADHS in February 1981 to purchase the necessary 1-square-mile parcel of Federal land for the HWF. ADHS approached the U.S. Bureau of Land Management (BLM) to buy the land, but BLM determined that the building of a HWF could significantly affect the quality of the human environment (U.S. EPA, 1983). Citing the National Environmental Policy Act of 1969, BLM announced that an EIS would be required before the sale could take place. BLM asked the U.S. Environmental Protection Agency (EPA) to prepare the EIS, because of EPA's greater expertise in the field of hazardous waste management (U.S. EPA, 1983).

In December 1981, ADHS announced the formation of a Hazardous Waste Facility Technical Advisory Group (HWFTAG) to assist in the development of the state hazardous waste management facility. The group consisted of representatives from public interest groups, private industry, and public agencies. The first author was invited to sit on HWFTAG's Public Involvement Coordination Subcommittee, and encouraged to develop a community survey to be cosponsored by Arizona State University and ADHS.

A meeting was sponsored by EPA in February 1982 to discuss the proposed HWF with residents at a local school. Residents expressed their anger over the decision-making process and pressed for answers to their questions about health and safety. Residents begrudgingly broke into small groups run by members from the League of Women's Voters to identify specific concerns and questions. No issues were resolved at the meeting and the relationship between local residents and government officials remained strained.

Survey Development and Objectives

Planning for the community survey began in earnest after the February 1982 meeting. The authors met with ADHS personnel regarding the format and focus of the survey, and followed up these meetings with an official proposal. Comments on the proposal by ADHS staff were insightful because of their different emphases. One emphasis was scientific, stressing the need for adequate questioning of the individuals' health history and their current symptomatology. Another emphasis was more political, focusing on the wording of individual questions in order to minimize the potential negative impact of the survey.

The general aim of the survey was to assess the psychological impact of the proposed HWF on residents living in the Rainbow Valley area. The objectives were to (a) identify the concerns of local residents; (b) assess the magnitude of psychological distress in the community; (c) determine whether residents who were involved in activities regarding the proposed facility were representative of the community at large; (d) suggest interventions to ameliorate the effects of stress; (e) serve as a vehicle to improve relations between the community and state government; and (f) generate baseline data for future studies aimed at determining whether or not health and quality of life had changed appreciably over time.

Gaining Entry Into the Community

After a draft of the survey had been developed, the first author spent 2 weeks meeting with identified community leaders in their homes. The purpose of these meetings was to explain the reason for the survey, to learn more about the concerns of the local population, and to have residents review a draft of the survey instrument. Most of the time was spent listening to local residents, who expressed their anger, concerns, and frustration regarding the proposed HWF. An effort was made to convey that their concerns were legitimate, their views were respected, and that their input was desired. Many residents were skeptical of the survey due to the involvement of ADHS; they pointed out their past negative experiences in working with the state government.

These meetings helped establish rapport with local residents and were successful for a number of reasons. First, the meetings were held in the homes of local residents. This placed residents at ease because they were in a comfortable environment and facilitated an informal discussion on a topic that bothered many people. Second, a state government liaison came to meet with them. This very act sent an implicit message to residents that they were respected and that their cooperation was valued. Third, the meetings were small, allowing for full participation among those in attendance and enabling residents who might have felt uncomfortable in larger settings to speak their mind. Fourth, their views were accepted and not challenged. In turn, the residents revealed valuable information about the community and past efforts to stop the proposed HWF from being built. They emphasized the need to make clear in all publicity to local residents that participation in the survey did not imply support of the proposed HWF.

Work on the survey intensified at ADHS after the Directors of the Division of Environmental Health and the Division of Disease Control gave their approval of the study proposal and received verbal approval of the project from the Director of ADHS. Interviewers were trained, sampling procedures outlined, and the search for a comparable control community begun. A special map of the sparsely populated Rainbow Valley area was developed, the preliminary survey instrument was field tested, and an article was submitted to the local newspaper to publicize the upcoming survey.

ADHS PULL-OUT AND REVISION OF THE SURVEY

On July 23, 1982, the survey came to a standstill, 1 week before it was to begin. The Director of ADHS in a single stroke cancelled the entire project and withdrew all resources. Explaining his decision, he stated that the health department would have to go back to the State Legislature for additional approvals in January 1983 and was worried that data from the survey might be used to force a delay in the State's much needed HWF. Political concerns once again had muddied the waters around Rainbow Valley.

As a consequence of the withdrawal of ADHS resources, the scope of the study was scaled down. The control community was eliminated and certain physical health questions were deleted from the interview schedule. New interviewers had to be recruited and trained, and separate funding obtained to cover interviewer costs and travel expenses. Nevertheless, 3 weeks later the survey began.

Sample

The sample consisted of 99 residents selected from Rainbow Valley, Arizona, one of two ways. The majority of the respondents ($n = 70$) were adults from randomly selected households in the community. A smaller group of people ($n = 29$) were residents who had attended meetings or voiced their opinion on public record about the proposed hazardous waste facility. This latter group was included to ensure a sizeable number of "involved" residents, so that questions could be answered regarding how representative these residents were of the community at large. Only one adult per household was interviewed.

A special map of sparsely populated Rainbow Valley was developed, because there was no acceptable alternative method of accurately identifying households in the area. The map identified all structures in the community, with each structure being given a separate number. In order to ensure the accuracy of the maps and to aid the interviewers in identifying households, all structures ($n = 398$) were included, whether or not they appeared inhabited. A random numbers list was used to identify households from which adults were to be interviewed.

Involved was defined as a local resident who could be identified from the attendance roster as having attended certain meetings or whose comments were documented in the January 1981 ADHS siting report (ADHS, 1981). A total of 65 involved residents were identified. Each name was assigned a number, and a random numbers list was used to identify persons to be interviewed.

Interviewers and Interview Training

Interviews were conducted by the first author, an undergraduate research assistant, a graduate student in clinical psychology, and a public health nurse. Interview training was conducted by the first author and included how to gain entry, establish rapport, and handle a variety of field problems. During training each interviewer practiced administering the instrument and role-playing a respondent, in order to sensitize the interviewer to both sides of the interview process. Two interviewers, the first author and the public health nurse, received additional training by field testing a preliminary form of the instrument. Because the first author completed 43% of the interviews, there was concern about possible interviewer bias. Analyses of variance on selected measures revealed no significant differences between interviewers.

Procedure

Interviewers were randomly assigned a list of households to survey. At households of involved residents, they were given the name of the person with whom to speak. At randomly selected households, procedures were employed to obtain proportionally equal numbers of male and female interviews. At the first household, the interviewer asked to speak to the head of the household (i.e., the man or woman of the house) that corresponded to the sex of the person first greeting the interviewer. Thereafter, the sex of the respondent was alternated.

Two callbacks were made in order to complete the interview before contacting residents of another household. Time and day of canvassing was varied to increase the likelihood of contacting residents and to interview them at convenient times.

Standardized interviewer instructions on how to greet each respondent, explain the purpose of the survey, and introduce different sections of the interview were included on the survey instrument. Most of the survey was administered orally, although one section was filled out by the respondent. The decision to have a self-administered portion of the interview was reached following the field testing of a preliminary interview form, in which all questions were orally administered. It was believed that more honest and valid responses could be obtained if certain personal and apparently sensitive questions were filled out by the respondent.

All interviews were conducted at the home of the local resident, and a great effort was made to interview respondents in private, away from children and other adults.

The Survey Instrument

The interview consisted of 187 open- and closed-ended questions, and included established standardized scales as well as scales developed specifically for this survey. Local residents were asked about their major concerns regarding the planned hazardous waste facility, their perception of the facility and potential problems, and what compensation they would like to receive for having the facility built in their community. An analysis of these items is the focus of the discussion to follow. The survey instrument also included an abbreviated form of the Ways of Coping Scale (Lazarus & Folkman, 1984), measures of self-efficacy (Pearlin & Schooler, 1978) and self-esteem (Rosenberg, 1965), and specially developed scales to assess sense of community and community involvement concerning the planned hazardous waste facility. Detailed information about these scales and their psychometric properties have been presented elsewhere (Bachrach & Zautra, 1985).

A 26-item self-administered demoralization scale was used to measure psychological distress. It was the same form of the instrument used by Dohrenwend et al. (Report of the Task Group on Behavioral Effects, 1979) in their assessment following the nuclear accident at Three Mile Island. Demoralization refers to psychological symptoms and reactions a person is likely to develop when faced with a serious predicament, for which the person can see no resolution (Frank, 1973). The scale measures concepts such as helplessness–hopelessness, dread, confused thinking, sadness, psychophysiological symptoms, and perceived physical health problems. Demoralization can be associated with both severe environmental stressors such as combat, natural disaster, chronic physical illness, and also diagnosed psychological disorders (Dohrenwend, Shrout, Egri, & Mendelsohn, 1980).

Characteristics of the Surveyed Population

There were 142 households or residents contacted and 99 interviews completed. The replacement of households or individuals accounted for 34 of the 43 non-completed interviews. The most common reason for replacement was that the residence was obviously vacant and not inhabited ($n = 17$), with examples being homes under construction, storage sheds, and abandoned trailers. Replacement also occurred if people (a) had lived less than 6 months in the community; (b) could not be contacted after two callbacks; (c) had moved out of the area; and (d) could not be located.

Only nine respondents refused to be interviewed, which corresponds to a 91.7% completion rate for residents who were eligible to be interviewed and could be located. All but one refusal came from the randomly selected sample, and as a group they tended to be fairly old, with an average age of 55 years.

The demographic characteristics of the 99 respondents who completed the interview are presented in Table 5.1. Most of the residents interviewed were Caucasian (82%), married (84%), and high school graduates (73%). They had lived an average of 8 years in the area, and slightly more than half were female (52%). The average age was 41 years, but there was a fairly equal distribution across age groups. The median family income was $21,000 with one quarter of the families earning under $13,000 and one quarter earning over $28,000. Sixty-three percent were employed full- or part-time, 7% were retired, 19% were homemakers, and 8% were unemployed. The median time per completed interview was 35 minutes.

Perceptions and Concerns of Local Residents Regarding the Proposed Facility

Eighty-five percent of the sample reported they were either concerned (34%) or upset (51%) when they learned the hazardous waste facility might be built. They perceived the facility as a threat, in contrast to the 15% who were either pleased (2%) or did not care (13%). A detailed analysis of how the "threat" sample coped with the planned HWF has been presented elsewhere (Bachrach & Zautra, 1985).

Most people believed the decision to select the Rainbow Valley site was either unfair (33%) or very unfair (49%), and viewed the Governor, the State Legislature, and certain other politicians as responsible for the decision. Half of those interviewed were either angry (20%) or very angry (30%) about the decision and generally did not see themselves becoming less angry over time. Still, over half of those interviewed (54%) saw themselves accepting the fact that the facility may be built, and two-thirds of the sample believed that it was either likely (18%) or very likely (44%) that the facility would be built.

Near the start of the interview people were asked the question, "Specifically, what are your feelings and opinions about the proposed facility?", and up to four responses per individual were recorded. Many responses fell into similar categories and are presented in Table 5.2. People were most concerned about water contamination, felt better sites were available for a hazardous waste facility in more isolated areas, and were worried about health hazards and transportation accidents. A series of closed-ended questions suggested that residents viewed the facility as a threat, were distrustful of those who may manage the facility, and believed few community benefits would come from it. Examples to support each of these statements are presented in Table 5.3.

TABLE 5.1
Demographic Characteristics of Surveyed Residents

Variables	Total (n = 99)	Random (n = 70)	Involved (n = 29)
Income			
Less than $3,000	3.0%	2.9%	3.4%
$3,000–$8,000	7.1	7.1	6.9
$8,000–$13,000	14.1	10.0	24.1
$13,000–$18,000	16.2	18.6	10.3
$18,000–$23,000	20.2	17.1	27.6
$23,000–$28,000	8.1	7.1	10.3
Over $28,000	24.2	28.6	13.8
Refused/Missing	7.1	8.6	3.4
Education			
Less than eighth grade	9.1%	10.0%	6.9%
Eighth or ninth grade	6.1	5.7	6.9
Some high school	12.1	14.3	6.9
High school graduate	38.4	35.7	44.8
Some college	24.2	24.3	24.1
College graduate	8.1	7.1	10.3
Post-graduate	2.0	2.9	0.0
Sex			
Male	48.5%	45.7%	55.2%
Female	51.5	54.3	44.8
Race			
White	81.8%	84.3%	75.9%
Mexican–American	10.1	12.9	3.4
Black	7.1	1.4	20.7
Other	1.0	1.4	0.0
Martial Status			
Married	83.8%	85.7%	79.3%
Single	7.1	8.6	3.4
Divorced	6.1	5.7	6.9
Widowed	3.0	0.0	10.3
Age			
18–24	10.1%	14.3%	0.0%
25–34	26.3	31.4	14.3
35–44	25.2	30.0	14.3
45–54	19.2	15.7	28.5
55–64	12.1	5.7	28.6
65 and over	6.1	2.9	14.3

TABLE 5.2
Major Concerns of Residents[a]

Percent	Response
43	Water contamination
33	Better/other sites/in more isolated areas/away from people
24	Health hazards
23	Transportation hazards/accidents
13	General safety concerns
12	Waste/chemical leakage
11	Don't like it/against it/here
10	Air pollution
8	Lower property value
7	Site selection a political decision
6	Need a site
6	Good/ideal site/if built right

[a]Based on the open-ended question, "Specifically, what are your feelings and opinions about the proposed facility?" Up to four responses per individual were recorded.

Residents also were asked what compensation or community benefits they felt would be most important. As can be seen in Table 5.4, the statements relating to health and safety were most important in the minds of local residents.

Comparing the Two Subsamples

The responses of the two subsamples were compared to determine whether people who attend meetings and publicly voice their opinion differ from residents selected at random from the community. The random and involved subsamples were significantly different in some respects but not in other ways. These reported differences were at the $p < .001$ level or less. The involved group was older (51 years vs. 37 years), had lived longer in the community (12 years vs. 7 years), and had known that the facility might be built for a longer period of time (2 years vs. 1 year).

Although most residents of both groups were opposed to the building of the facility, the involved group's opposition was more intense. They were "upset" when they learned the facility might be built, whereas the rest of those interviewed were more likely to respond "concerned." They believed the decision to select the Rainbow Valley site was "very unfair" whereas others viewed it as "unfair." Similarly, those actively involved were generally "angry" over the decision to build the facility compared to "somewhat angry" for the rest of the sample.

There were no significant differences between the two groups on many measures. Similar proportions of each group owned their homes and believed that property values would decrease. Both groups equally believed that the facility would probably be built and that good relations were not likely to develop

Table 5.3
Perceptions of Facility, Facility Management and Expected
Community Benefits

Perception Items	Agree or Agree Strongly	Disagree or Disagree Strongly	Don't Know
(1) FACILITY VIEWED AS A THREAT			
The likelihood of an accident occurring is extremely slight	16%	73%	11%
There will be a high potential for ground or surface water contamination.	81%	6%	12%
People will face a greater risk of becoming sick	69%	18%	13%
The facility represents a threat to my well-being	70%	22%	8%
(2) DISTRUST OF FACILITY MANAGEMENT			
Public complaints on the facility's operation will be investigated promptly.	25%	65%	10%
The public and local community will be advised of changes in operations and types of waste handled at the facility	15%	78%	7%
A good working relationship will develop between local residents and the facility operator.	15%	72%	13%
(3) FEW COMMUNITY BENEFITS EXPECTED			
Fire and emergency systems in the community will be improved.	25%	55%	19%
Better medical facilities will be provided for the community	15%	69%	16%
The facility will generate tax dollars for the community	26%	52%	22%

between the local community and the facility operator. A very important finding was that both groups ranked the seven community benefit or compensation items in a similar fashion, viewing health and safety concerns as most important.

Psychological Distress in the Community

The demoralization scale has been used previously to estimate generalized distress following the nuclear accident at TMI (Dohrenwend et al., 1981). That study found that shortly after the accident 26% of the local population scored above the mean of clients of community mental health centers, most of whom

TABLE 5.4
Desired Compensation in Order of Importance[a]

Item	Average Ranking
The health of residents should be regularly monitored.	2.51
Better medical facilities should be provided for the community.	3.09
Fire and emergency systems in the community should be improved.	3.20
Some type of compensation should be provided to residents affected by the facility.	3.66
Access roads should be improved (e.g., paved) and widened.	4.61
Employment opportunities for local residents should be provided.	5.43
The facility should generate tax dollars for the community.	5.44

[a]Based on rank ordering of items, with "*1*" being most important.

were suffering from chronic mental disorders. Using similar mean scores as employed in the TMI study (greater than 25 and 30 for men and women, respectively), it was found that 34.3% of the Rainbow Valley sample scored above the mean. The average score for this highly demoralized group was 39.0 compared to 15.5 for the rest of the sample. The scale was quite reliable, having an internal consistency of .92, as measured by coefficient alpha.

Analyses are still being conducted to better understand which residents are most at risk, but certain significant findings are worth noting. Demographically, highly demoralized residents had lower family incomes and tended ($p < .07$) to be younger than residents not so distressed. They had a lower sense of self-efficacy and self-esteem, focused more on their emotional reactions to the stressor and were less likely to be involved in activities concerning the hazardous waste facility. In fact, a causal chain is implied in which decreased self-efficacy promoted increased emotion-focused coping, which contributed to less community involvement (see Bachrach & Zautra, 1985, for fuller discussion of these findings).

DISCUSSION

The results paint a strong adversarial relationship between the community and government institutions. The community does not trust the state government as evidenced by perceptions that (a) politicians were responsible for the decision to select the Rainbow Valley site; (b) the decision was unfair; and (c) facility management cannot be trusted. Furthermore, residents viewed the facility as a threat to their well-being and expected few community benefits. In a nutshell, the community has said that it has little to gain and potentially much to lose as a result of this facility.

Even though local residents did not want the facility, the majority of them believed that it would be built in their community. This suggests that residents did not believe that they could do much about the present situation, which may promote a sense of powerlessness and alienation among them.

A surprisingly large proportion of the residents surveyed scored above the mean of community mental health clients on the demoralization scale. A comparison with data gathered near TMI may help place this finding in perspective. Shortly after the TMI nuclear accident, 26% of the local population scored above the mean, while in subsequent months the rate dropped to 15% or less (Dohrenwend et al., 1981). In Rainbow Valley, 34% of the sample scored above these mean scores. Although the normative data collected in the TMI area may not be appropriate for comparison among residents of the present investigation, it does suggest that a considerable number of residents are severely demoralized.

Highly demoralized residents were less likely to be involved in community activities. As a consequence, the psychologically most fragile and vulnerable residents were underrepresented in community activities and thus less visible to government officials attempting to "take the pulse" and assess the impact of the HWF on the local population. This suggests that intervention programs aimed at lowering distress need to go directly to the local residents and not have residents come to them. The usual mode of service delivery, in which the patient seeks out the health professional, will likely not attract those persons in greatest need.

The proposed hazardous waste facility undoubtedly was not the sole reason for the distress many people were experiencing; however, it was likely a contributing factor. Members of the local community voiced numerous concerns about safety, water contamination, and health. The residents seemed distrustful of the people who may manage the plant and viewed the facility as a threat. To truly understand the impact of this proposed facility would require periodic reassessments of the local population, along with assessments of a similar community to serve as a control group.

A continuation of this adversarial relationship would not be beneficial to any of the parties involved. If the community does not feel it can trust the State or the facility operator, distress among local residents may be increased once the building of the hazardous waste facility begins; for if residents cannot trust the party monitoring the safety of the facility, to whom can they turn? The state government has little to gain by having its credibility questioned continuously by local residents. Furthermore, such criticism does not remain a local issue but spreads via the media. Similarly, the facility operator will likely be backed into a corner, having to frequently defend its actions to people looking for mistakes.

A reconciliation among the parties will not be easily achieved. The local community has fought hard to stop the building of the facility and 46% of the residents still do not see themselves accepting the fact that the hazardous waste facility will be built. There is no simple solution, but a dialogue between representatives of the State and local community would be a step in the right direction. Many issues must be sincerely dealt with and resolved if relationships are to improve. An overriding issue may be one of justice and fairness. Few people want a hazardous waste facility located near them. Should those who must assume more than their fair share of the social cost of hazardous waste production receive something in return? It is interesting to note that local residents viewed as most important the monitoring of their health and better medical facilities, not monetary compensation or employment opportunities.

THE ROLE OF PSYCHOLOGY

Most people in the EIS process have backgrounds in engineering, chemistry, geology, hydrology, toxicology, and public administration. This technical expertise is necessary, but so is the ability to work effectively with groups and individuals, which is where psychologists have expertise. Because the EIS is mandated when a federal project could significantly affect the quality of the human environment, the needs and concerns of people living in the surrounding environment should not be ignored.

A major issue from the start was the residents' lack of trust in the state and federal government to protect their health and safety. State government and EPA officials held meetings to try to allay these fears, but were not very effective. The audiences generally became angry and hostile rather than relieved.

Part of this anger was simple frustration due to poor communication. Government officials often sincerely tried to answer questions, but they failed to grasp the dynamics of the group process at these meetings. The government officials addressed the questions intellectually, ignoring the emotional issues. They attempted to correct their audience's supposed misconceptions through education without legitimitizing the many fears and concerns of the residents. The residents spoke on a more emotional level. Although they, likely, would not have been completely satisfied with any answer, they became increasingly angry when state officials failed to acknowledge their fears. Government officials dealt with this hostility often by appearing defensive and paternalistic, which only enraged these residents more. Such interactions helped perpetuate the strained relations between local residents and government officials.

A psychological consultant could have been used to (a) serve as a facilitator at these meetings; (b) train government officials in human relation skills and in how to deal with a hostile audience; and (c) suggest alternative ways of meeting with local residents other than in large community forums. For example, meeting with local residents in small groups in their homes, as the first author did when developing the survey, may have been more beneficial and cost-effective.

A step in the right direction was taken when the state recognized and validated the concerns of residents. The final EIS included the results of this survey along with the following comments:

Partly in response to [this] study, the Director of ADHS has committed the Department to gathering baseline health of area residents throughout construction and operation of the facility. This health monitoring program, together with the permit conditions, may help to address the concerns expressed by survey participants regarding the potential health risks associated with the facility. (U.S. EPA, 1983, p. 59)

A major reason why ADHS officials and community residents had difficulty understanding one another is because they likely had very different perceptions of the risks associated with hazardous waste. Slovic, Fischhoff, and Lichtenstein (1980) had experts and members of the lay public rate 30 technologies and activities (e.g., smoking, x-rays, nuclear power) on nine qualitative characteristics such as voluntariness, controllability, familiarity, catastrophe potential, and dread. They found that the lay judgments of risk could be predicted to a great extent from ratings of perceived dread, severity (likelihood of fatality), subjective fatality estimates in a given year, and potential for disaster. In contrast, the experts' judgments of risk were not related to any of the nine qualitative characteristics.

The catastrophic potential of hazardous waste may be the most important factor in understanding the local community's fear of, and opposition to, the planned facility (Slovic et al., 1980). Even if the probability of an accident is low, the consequences of an accident are extremely severe. This problem has plagued the nuclear power industry, where annual fatalities have been low compared to other hazards, but where the potential of a major disaster remains. Furthermore, it is difficult to educate people about the safety of a hazardous waste facility when, almost daily, new problems involving hazardous waste disposal are reported by the media.

The long-term psychological impact of living near hazardous waste facilities is not known. Those residents who perceived the risk as too great will likely move away; but others with similar perceptions may remain because of economic necessity or family ties. A recent appeals court ruling stated that psychological stress should be considered in the EIS process when (a) it is severe enough to threaten physical health and (b) it can be reliably measured (Hartsough & Savitsky, 1984). Compliance with these legal guidelines requires a better understanding of how people perceive and cope with environmental hazards. Longitudinal studies that allow for multiple assessments before, during, and after the environmental change, and that include matched control groups, hold the most promise for increasing our knowledge base. Still, much can be done presently to avoid or minimize the impact of stress by better understanding the concerns and fears of local residents, and responding sensibly to their needs. This study showed that many of the human problems resulting in stress seemed to involve communication, decision-making, and perceptual processes. These are areas where psychologists have special skills and training.

REFERENCES

Arizona Department of Health Services (1981, January). *Report to the Arizona State Legislature regarding siting of a statewide hazardous waste disposal facility.* Phoenix: Author.

Bachrach, K. M., & Zautra, A. J. (1985). Coping with a community stressor: The threat of a hazardous waste facility. *Journal of Health and Social Behavior, 26,* 127–141.

Bromet, E., Parkinson, D., Schulberg, H. C., Dunn, L., & Gondek, P. C. (1980). *Three Mile Island: Mental health findings.* Washington, DC: National Institute of Mental Health.

Dohrenwend, B. P., Dohrenwend, B. S., Warheit, G. J., Bartlett, G. S., Goldsteen, R. L., Goldsteen, K., & Martin, J. L. (1981). Stress in the community: A report to the President's Commission on the Accident at Three Mile Island. *Annals of the New York Academy of Sciences, 365,* 159–174.

Dohrenwend, B. P., Shrout, P. E., Egri, G., & Mendelsohn, F. S. (1980). Non-specific psychological distress and other dimensions of psychopathology. *Archives of General Psychiatry, 37,* 1229–1236.

Epstein, S. S., Brown, L. O., & Pope, C. (1982). *Hazardous waste in America.* San Francisco: Sierra Club Books.

Frank, J. D. (1973). *Persuasion and healing.* Baltimore: The Johns Hopkins Press.

Hartsough, D. M., & Savitsky, J. C. (1984). Three Mile Island: Psychology and environmental policy at a crossroads. *American Psychologist, 39,* 1113–1122.

Kolbe, J. (1981, January 22). Rainbow valley: The saga of a hazardous waste dump site. *Phoenix Gazette,* p. 8.

Lazarus, R. S., & Folkman, S. (1984). *Stress, appraisal, and coping.* New York: Springer.

Metropolitan Edison Co. v. People Against Nuclear Energy, 75L. Ed. 2nd, 534 (1983).

Pearlin, L. I., & Schooler, D. (1978). The structure of coping. *Journal of Health and Social Behavior, 19,* 2–21.

Report of the Task Group on Behavioral Effects. (1979). *Presidents' Commission on the Accident at Three Mile Island* (Report GPO No. 052500300732-1). Washington, DC: U.S. Government Printing Office.

Rosenberg, M. (1965). *Society and the adolescent self-image.* Princeton, NJ: Princeton University Press.

Slovic, P., Fischhoff, B., & Lichtenstein, S. (1980). Facts and fears: Understanding perceived risk. In R. C. Schwing & W. A. Albers, Jr. (Eds.), *Societal risk assessment: How safe is safe enough?* (pp. 181–214). New York: Plenum Press.

U.S. Environmental Protection Agency (1979, November). *Siting of hazardous waste management facilities and public opposition* (SW-809). Washington, DC: U.S. Government Printing Office.

U.S. Environmental Protection Agency (1983, July). *Final environmental impact statement for proposed Arizona hazardous waste management facility.* San Francisco: Author.

6

Predictors of Psychological Distress in the Community Following Two Toxic Chemical Incidents

Jeffrey S. Markowitz
Columbia University

Elane M. Gutterman
Mount Sinai School of Medicine

The chemical disaster in Bhopal, India has heightened the world's awareness of the potential dangers associated with the production, storage, and use of toxic materials. This incident resulted in immediate and widespread tragedy and death. More commonly, however, toxic exposure events do not result in such extreme negative health outcomes. Individuals may experience transient or chronic symptoms depending on the intensity, toxicity, and duration of the exposure. In addition, members of "exposed" communities often are faced with the long-term threat that at some point in the future their health may be undermined as a result of that exposure. The immediate psychological effects of exposure, plus living with these prolonged risks, is not yet well understood either on a short- or long-term basis.

Most of the literature describing levels of psychological distress or psychophysiological symptoms following a hazardous exposure incident focuses on the Three Mile Island (TMI) nuclear accident. Groups from TMI targeted for study include: random samples of adult heads of household in the general population (Dohrenwend et al., 1981; Fleming, Baum, Gisriel, & Gatchel, 1982), psychiatric out patients (Bromet, Schulberg, & Dunn, 1982), and workers at the nuclear plant (Kasl, Chisholm, Eskenazi, 1981). In the general population, in the months immediately following the accident, the highest levels of distress were reported by those living within 5 miles of the plant. In addition, although women invariably rated themselves as more distressed, currently married men and women were less distressed relative to those not currently married (Dohrenwend et al.,

1981). More than a year following the accident, a small community sample located within 5 miles of the damaged TMI plant was compared with community samples who faced no similar threat. Psychophysical symptoms were higher among TMI residents, relative to controls, with those having lower social support being additionally impaired (Fleming et al., 1982).

In regard to special groups, patients treated at community mental health centers serving the TMI area were compared with those similarly treated in western Pennsylvania. Although no differences were found between the two groups in self-reported levels of psychophysical symptoms, among TMI patients three factors distinguished between the high- and low-symptom groups. These were (a) the belief that TMI was currently dangerous, (b) lower family incomes, and (c) poorer general social support (Bromet et al., 1982).

Nuclear workers at TMI were compared to workers at Peach Bottom, a nuclear plant located in the same general region of Pennsylvania. When broken down by supervisory status, both supervisory and non-supervisory workers at TMI reported higher levels of psychophysiological symptoms during the time of the accident. In contrast, levels of psychological distress were higher among non-supervisory workers at TMI, whereas no difference in distress was found between supervisors at the two sites (Kasl et al., 1981).

Asbestos-exposed workers also have been studied in regard to psychological dimensions and compared to postal workers (Lebovits, 1984). No difference was found between the two groups on current psychopathological symptoms in the past month. Nevertheless, relative to postal workers, asbestos workers felt that they were significantly more likely to develop cancer.

Several points can be made regarding this existing literature on psychological ramifications of hazardous exposure incidents:

1. Pioneering work at TMI needs to be extended to a range of hazardous situations in order to identify similarities and differences between types of incidents.

2. In developing a core of critical dimensions to be investigated in hazardous exposures, results need to describe both factors that were and were not predictive of distress.

3. Further exploration of attitudinal predictors of psychological distress is advised. Although sociodemographic characteristics such as age, sex, and marital status may identify population subgroups at risk for higher psychological distress, they are not aspects that, practically speaking, can be modified. In contrast, attitudinal dimensions, such as the level of perceived health threat, may be influenced by situational factors, for example clear communication of risks by environmental and health officials and access to subsidized health services to check on possible physical sequelae of exposure. As noted previously, Bromet et al. (1982) found that the perception of TMI as currently dangerous differentiated between TMI area mental health patients reporting high versus low distress.

4. The time interval following the incident must be clearly identified. In

addition, great effort must be made to obtain data in the weeks immediately following the incident, rather than relying on retrospective reports of distress.

In this chapter, we examine predictors of psychological distress in the community following two toxic chemical incidents. The first incident occurred in April 1980 in Elizabeth, New Jersey and involved an explosion of toxic chemical wastes. Residents from a community several miles away in Staten Island, New York comprised the target population. The second incident occurred more than 4 years later in Linden, New Jersey and involved leaking malathion pesticide fumes. The population of interest in this case was residents from the nearby New Jersey communities of Linden and Perth Amboy.

An objective of the Staten Island study was to uncover which demographic, situational, attitudinal, and psychological symptom-related variables best predicted post-incident psychological distress. One such predictor was perceptions of threat to physical health (Markowitz & Gutterman, 1984). In this chapter, we report our findings in respect to predictors of psychological distress in the Staten Island community study and also attempt to replicate these results using the Linden–Perth Amboy community sample.

OVERVIEW OF THE TOXIC EXPOSURE INCIDENTS

The two incidents that are the focus of this chapter occurred in Elizabeth and Linden, New Jersey. The sites of the two incidents are far enough apart to enable us to safely assume that none of our respondents were affected by both events. In addition to the distance that separates the two communities, prevailing wind currents make it nearly impossible that fumes from either of the incidents came close to the other community. The two incidents are described in turn.

At 11:00 P.M. on April 21, 1980, a toxic chemical waste incident occurred at the Chemical Control facility in Elizabeth, New Jersey. About 40,000 highly flammable, 55 gallon drums of toxic industrial wastes exploded into the air sending a huge dark plume of chemical fumes over parts of nearby Staten Island, New York. The following day, schools were ordered closed and people were urged to stay indoors with their windows shut. Evacuation plans were drawn up and several major highways and an important bridge were closed. (Pfeiffer, Patrick, & Cross, Staten Island Advance, 1980)

The contents of the chemical clouds that blew over parts of Staten Island were never clearly identified. Two days after the incident, however, state officials ordered a ban on fishing in the New York harbor based on the identification of some 14 toxin substances seeping into the Arthur Kill and Elizabeth Rivers (Pfeiffer, Staten Island Advance, 1980) These chemicals had leaked into these waterways from the smoldering wreckage of the Chemical Control facility.

Eleven of these 14 chemicals were included in a report to the U.S. Senate issued in 1980 (U.S. Senate, 1980). In this report, as shown in Table 6.1,

TABLE 6.1
Published Findings Related to Chemicals Involved in the Elizabeth
Incident

	Muta-genicity	Carcino-genicity	Terato-gencity	Feto-toxicity	Neuro-toxicity	Hepato-toxicity	Renal-toxicity
Benzene	+	+	+	+	?	?	?
Chloroform	−	+	+	+	?	+	?
Dichloroethane	−	+	−	+	+	?	?
Dichloromethane	+	−	−	−	+	?	?
Ethel Benzene	?	−	?	?	?	?	?
Tetrachlorethylene	−	+	−	−	+	+	+
Trichlorethane	+	−	−	−	?	?	?
Trichlorethylene	+	+	−	−	+	+	?
Toluene	+	−	−	+	+	?	?
Xylene	?	?	?	?	+	?	?
Polychlorinated Biphenyls (PCBs)	+	+	+	+	+	+	?

+ = At least one report of positive listed effect
− = Only negative data
? = No data in the literature
Adapted and modified from Epstein, Brown, and Pope (1982)

different chemicals were rated in seven categories of toxicity with respect to what is known about them in the published literature. As can be seen, these chemicals most often were found to be positive for mutagenicity, carcinogenicity and neurotoxicity. Table 6.1 depicts the many gaps in our knowledge of possible adverse physical health consequences resulting from chemical exposure to specific agents. However, our knowledge base is even less developed when it comes to answering questions regarding how these chemicals may affect us psychologically—either through some direct biophysiological process or indirectly through stress.

Beginning a little over a month following the toxic chemical fire in Elizabeth and continuing for about 10 weeks, we interviewed a random sample of 67 community residents from a nearby exposed area on Staten Island called Travis. Using a modified version of the assessment battery used by the Behavioral Task Force of the President's Commission on the accident at Three Mile Island (Dohrenwend et al. 1981), Travis residents were questioned on a range of demographic, attitudinal, situational, and psychological symptom-related dimensions.

On October 6, 1984, a malathion pesticide incident occurred at the American Cyanamid plant in Linden, New Jersey. An outdoor tank filled with malathion overheated and a yellow cloud of toxic chemical smoke escaped into the environment, particularly in the vicinity of the Arthur Kill River that separates Staten Island from New Jersey. Nearly 200 persons visited local hospitals in New Jersey and Staten Island complaining of symptoms associated with the incident. In addition, police and other public officials received thousands of calls from members of the community who were experiencing symptoms and were concerned about them (McFadden, 1984).

An entire crew of seamen docked aboard a commercial ship located less than 1,000 feet from the site of the Linden incident were examined at local hospitals for symptoms associated with the incident. Many of these seamen were experiencing a range of physical symptoms such as nausea, skin irritation, dizziness, and headaches. In a systematic study of 22 of these seamen and 21 seamen controls, subjects were higher on overall post-incident physical symptomatology. The subjects were significantly higher than controls on many physical symptoms specifically associated with malathion exposure (Markowitz, Gutterman, & Link, in press).

The study of the seamen was a component of a larger community study undertaken in Linden and Perth Amboy. A major objective of the community study was to attempt to develop reliable and valid measures that would be useful in assessing psychosocial response to toxic exposure events. The new instrument tested in this research is called the Toxic Exposure Research Interview (TERI) and includes an extensive battery of pre- and post-incident physical and psychological symptom-level measures.

Beginning only 2 weeks following the malathion incident and continuing for the next 8 weeks, mail questionnaires were completed by 65 Linden–Perth Amboy heads of households. The research tool used in this study was derived from the instrument used in the Travis study and included a demoralization scale and a measure of perceptions of threat to physical health. Other measures used in Linden–Perth Amboy included: pre-incident level of psychological symptomatology, attitudes toward moving, and distrust in authorities.

THE SAMPLES

Travis

Our sample was drawn from an official Staten Island City Hall document that included the addresses of all Travis residents. Using strict probability sampling techniques, 85 residential households were randomly selected and 67 interviews completed with adult heads of households. Face to face interviews were conducted in subjects' homes.

Linden and Perth Amboy

Our sampling frame included all adult heads of Linden and Perth Amboy households who had telephones and were listed in their respective telephone directories. Randomly chosen were 120 names or 60 from each community taken from the telephone directory. Instead of using face to face interviews, mail questionnaires were employed and in all, 65 completed surveys were returned.

DESCRIPTION OF THE MEASURES

Post-Incident Psychological Distress

Demoralization is a feeling state that a person may experience when they feel trapped by circumstances and see "no way out" (Frank, 1973). Clinically, demoralization may be associated with symptoms of depression and sadness, deficits in self-esteem, feelings of impending dread, feelings of hopelessness and helplessness, as well as somatic-type symptoms.

An operational measure of demoralization has been developed and tested by Dohrenwend and colleagues (Dohrenwend et al., 1981) and is part of the Psychiatric Epidemiology Research Interview (PERI). The PERI has been used extensively in epidemiological studies as a screening instrument to detect untreated psychiatric disorder (Dohrenwend, Shrout, Egri, & Mendelsohn, 1980). A modified version of this multi-item scale was used by the President's Commission at Three Mile Island and can be viewed as a measure of non-specific psychological distress. The modified version of the measure included 26 items and correlated .98 with the original composite scale formed from the larger set of demoralization items. The Travis and Linden–Perth Amboy studies respectively utilized 24 and all 27 demoralization items.

All of the demoralization questions in the Travis and Linden–Perth Amboy studies were prefaced with the phrase "Since the incident in (Elizabeth or Linden), how often have you felt. . . ." For most of the items, response categories include never, almost never, sometimes, fairly often, and very often. A list of the items included in the PERI demoralization scale are shown in Table 6.2. The scoring of PERI demoralization scale scores range from 0 to 4 with higher scores reflecting higher levels of demoralization.

Number of Psychological Symptoms Before the Incident

Meaningful predictors of demoralization are best identified when there is baseline data on levels of psychological symptomatology. This was accomplished by using a count of the number of psychological symptoms reported retrospectively for a period of a month preceding the incidents in Elizabeth and Linden. This measure has been used by Goldsteen (1983) in longitudinal studies of residents living within 5 miles of TMI. The 22 psychological symptoms included in this measure are shown in Table 6.3.

TABLE 6.2
27 Item PERI Demoralization Scale

 1. Feelings of sadness or depression—feeling blue
 2. Felt lonely
 3. Felt anxious
 4. Nervousness, being fidgety or tense
 5. Restlessness
 6. Worry type
 7. When angry, get headaches, stomach pains, cold sweats
 8. Acid or sour stomach
 9. Appetite been poor
10. Cold sweats
11. Headaches or pains in the head
12. All different kinds of ailments in different parts of your body
13. Nothing turns out for you the way you want it to
14. Completely helpless
15. Completely hopeless about everything
16. Feared going crazy, losing your mind
17. Attacks of sudden fear or panic
18. Feared that something terrible would happen to you
19. Confused and had trouble thinking
20. Trouble concentrating or keeping your mind on what you were doing
21. Confident
22. Uselessness
23. Failure in life
24. Much to be proud of
25. What grade would you give yourself
26. Satisfied with yourself
27. Wondering if anything is worthwhile anymore

This measure of pre-incident psychological distress is likely to covary with demoralization and is intended to control for symptoms existing prior to the event. The measure is simply an unweighted count of the items endorsed by the subject and theoretically, could range from 0 to 22.

Demographics

A wide range of demographic variables were used in both the Travis and Linden-Perth Amboy studies. They included (a) age; (b) sex; (c) high school graduate; (d) having a pre-teenaged child; (e) number of years lived in area; (f) marital status; and (g) homeowner status.

Attitudinal

Three attitudinal measures developed by TMI researchers for use in studies conducted by the Behavioral Task Force of the President's Commission at TMI were incorporated into the Travis and Linden–Perth Amboy studies. These are

TABLE 6.3
22 Items In Measure Of Psychological
Symptoms Before Incident

1. Strong feelings of fear
2. Periods of anger
3. Periods of extreme worry
4. Periods of extreme upset
5. Nausea
6. Stomach troubles
7. Headaches
8. Diarrhea
9. Constipation
10. Spells of crying
11. Loss of appetite
12. Increase in appetite
13. Trouble in sleeping
14. Sweating spells
15. Feeling trembly or shaken
16. Heart pounding or racing
17. Difficulty making decisions
18. Irritability
19. Nightmares or bad dreams
20. Feeling afraid in open spaces or in the street
21. Thoughts of ending your life
22. Feelings of being trapped

(a) perceived threat to physical health; (b) distrust in authorities; and (c) attitudes toward moving.

Perceived threat to physical health categorizes attitudes about being contaminated and whether one's health has deteriorated as a result of the incident. Distrust is a construct that taps attitudes pertaining to whether truthful information, associated with the exposure event, has been provided by federal, state, and/or local officials. Feelings about whether the incident has sparked a desire to move out of the area are identified in the attitudes toward move scale. All three variables may predict post-incident psychological distress.

Scale scores for all three attitudinal measures range from 1 to 3 with higher scores reflecting greater perceived threat, more distrust, and a greater desire to move out of the area.

RESULTS

Participating in the study were 132 subjects, including 67 from Travis and 65 from Linden–Perth Amboy. Of the 65 subjects in the latter sample, 35 were from Linden and 30 from Perth Amboy. The overall response rates in Travis and Linden–Perth Amboy were 86% and 74% respectively. Table 6.4 summarizes the field operations of the Travis and Linden–Perth Amboy studies.

To improve their internal consistency, several scale items from each of the attitudinal scales used in Travis had to be dropped from the Linden–Perth Amboy analyses. Items were omitted in accordance with objective criteria namely if substantial improvement in alpha resulted from dropping an item. Table 6.5 includes alpha reliability coefficients for the survey scales used in the two studies. The survey scales used in the two studies had acceptable alpha reliability coefficients ($\geq.67$) with the exception of attitudes toward moving in the Linden–Perth Amboy study. After dropping one item from the attitudes toward moving scale, the remaining two items had an alpha of .41.

The Linden and Perth Amboy community samples were similar across a range of study variables that included: age, sex, education, marital status, homeowner

TABLE 6.4
Summary of Travis and Linden-Perth Amboy Field Activities

	Travis	Linden-Perth Amboy
	N	N
a. Households Selected	85	120
b. Unavailable[a]	7	32
c. Refusals	9	23
d. Breakoffs	2	0
e. Interviews Completed	67	65

Response Rates for Travis and Linden-Perth Amboy = 86% and 74% respectively. This is computed using the following formulas:

$$\frac{(e)}{(a-b)}$$

[a]For example, moved, vacation, ill or died.

TABLE 6.5
Alpha Reliability Coefficients of Survey Scales

	Travis		Linden-Perth Amboy	
Scale	# Items	Alphas	# Items	Alphas
Demoralization	23	.93	27	.89
Psychological Symptoms Before Incident	22	.87	22	.79
Perceived Threat to Physical Health	3	.69	2	.78
Distrust in Authorities	5	.84	3	.67
Attitudes Toward Moving	3	.74	2	.41

status, date of interview, number of pre-incident psychological symptoms, demoralization following the malathion incident, levels of perceived threat to physical health, attitudes toward moving, and distrust in authorities. The Linden and Perth Amboy communities differed on distance from site of incident and number of years lived in the areas. The difference in distance from site was expected because the sampling strategy called for stratification by community to reflect distance from the toxic leak. The northern section of Perth Amboy is located about 8 miles from South Linden and about 12 miles from the American Cyanamid plant where the incident occurred. This stratification procedure is designed to answer the question of whether distance from site of incident affected levels of psychological distress. With regard to number of years lived in area, residents in Perth Amboy lived in their community about 10 years longer than residents in Linden. However, both groups are long-term residents of their respective communities, averaging 23 years in Linden and 33 years in Perth Amboy. Due to their overall similarities, the Linden and Perth Amboy samples were aggregated and jointly compared to the Travis sample in the analyses that follow. (See Table 6.6 for a complete summary of the similarities and differences between the Linden and Perth Amboy samples.)

In contrast to the similarities between the two communities involved in the malathion incident, the Travis sample differed in many respects from our combined Linden–Perth Amboy sample. These differences include: number of pre-incident psychological symptoms (Travis mean = 3.3 vs. 1.7 in Linden–Perth Amboy), sex (43% male in Travis vs. 60% in Linden–Perth Amboy), marital status (79% married in Travis and 61% in Linden–Perth Amboy), and homeowner status (94% owned their homes in Travis compared to 63% in Linden–Perth Amboy). Also, there was a trend for the two groups to be different on levels of

TABLE 6.6
Similarities and differences between Linden and Perth Amboy

Similarities	
Age ($p > .9$)	Distrust ($p > .3$)
Number Psychological Symptoms Before the Incident ($p > .6$)	Demoralization ($p > .9$)
Homeowner ($p > .7$)	Date of Interview ($p > .7$)
High School Graduate ($p > .9$)	Pre-teenaged Child ($p > .4$)
Sex ($p > .8$)	Marital Status ($p > .5$)
Perceived Threat to Physical Health ($p > .8$)	Attitudes Toward Moving ($p > .8$)

Differences	

Distance from Site of Incident (Perth Amboy is further $p < .001$). Years lived in area (Linden is lower $p < .05$).

demoralization following the respective incidents ($p > .05 < .1$) with Travis residents being higher (Travis demoralization mean was 0.9 compared to 0.7 in Linden–Perth Amboy). In addition, the two groups differed on perceived threat to physical health (Travis mean = 2.0 vs. 1.7 in Linden–Perth Amboy) and distrust (Travis mean = 2.8 vs. 2.3 in Linden–Perth Amboy). These results should be interpreted with caution. The threat and distrust scales used in Linden–Perth Amboy had fewer items than their respective scale counterparts used in Travis. Overlapping items, however, were identical. The communities were similar on age (mean age in Travis = 55 vs. 52 in Linden–Perth Amboy), having a pre-teenaged child (19% in Travis vs. 21% in Linden–Perth Amboy), years lived in area (mean number years in Travis = 32 vs. 27 in Linden–Perth Amboy), and attitudes toward moving (Travis mean = 2.3 vs. 2.2 in Linden–Perth Amboy). In addition, using graduated from high school as a cutoff, there was no significant difference between Travis (55% graduated) and Linden–Perth Amboy (59% graduated) on education. On the average, Travis residents lived 5.8 miles from the site of the Elizabeth incident and Linden–Perth Amboy residents lived 6.8 miles from the site of the malathion incident. This difference in miles lived from site of incident(s) was not significant. (See Table 6.7 for a statistical comparison of Travis and Linden–Perth Amboy.)

Holding constant the number of psychological symptoms before the incident, 12 independent study variables were entered individually in a series of multiple regression tests. The independent variables included the demographic, attitudinal, and situational measures described earlier. By removing the last preceding independent variable we were able to assess which of these were significant in adding to our prediction of demoralization scores. There were a total of five variables in Travis where significant F-change statistics were obtained. These were (a) having a pre-teen aged child; (b) date of interview; (c) distrust; (d) attitudes toward moving; and (e) perceived threat to physical health. In addition, at Travis, there were two variables where the p-value for the F-change test was $> .05$ and $< .1$ reflecting a trend for significance. These variables were education and number of years lived in area. All of the beta coefficients of the significant and trend-significant independent variables were positive with the exception of date of interview and number of years lived in area.

In Linden–Perth Amboy, significant F-change statistics, predicting demoralization, were obtained for two of the independent variables namely having a pre-teenaged child and perceived threat to physical health. In addition, the p-value for the F-change test for number of years lived in area was $> .05$ and 1. Together with pre-incident psychological symptom levels these variables accounted for about 56% of the variance in Linden–Perth Amboy demoralization scores. (See Table 6.8 for a complete description of the results of the F-change tests for both Travis and Linden–Perth Amboy.)

To identify the best predictors of demoralization in the Travis sample, the seven significant and trend-significant independent variables were entered into

TABLE 6.7
Comparison of Travis and Linden-Perth Amboy on Study Variables

Variables	TRAVIS (Mean or %)	LINDEN-PERTH AMBOY (Mean or %)	p-value
Demographics			
Mean age	55	52	NS
% Male	43	60	SIG*
% Homeowner	94	63	SIG**
% Married	79	61	SIG*
% High School Graduate	55	59	NS
% with Pre-teenaged Child	19	21	NS
Situational Factors			
Mean Distance From Site	5.8	6.8	NS
Mean Years Lived in Area	32	27	NS
Psychological Symptoms			
Mean# Pre-Incident	3.3	1.7	SIG***
Demoralization	0.9	0.7	TREND
Attitudes			
Mean Perceived Threat to Physical Health	2.0	1.7	SIG*
Mean Distrust in Authorities	2.8	2.3	SIG**
Mean Attitudes Toward Moving	2.3	2.2	NS

NS; Not Significant ($p > .05$)
TREND: Just Not Significant ($p > .05 < .10$)
SIG*; Significant ($p > .01 < .05$)
SIG**; Significant ($p > .001 < .01$)
SIG***; Significant ($p < .001$)

a multiple regression test simultaneously with levels of pre-incident psychological symptomatology. The following variables were significant in this analysis: attitudes toward moving, number of psychological symptoms before the incident, and perceived threat to physical health. Together, these variables accounted for 59% of the variance in demoralization scores. This model tested for more rigorous predictors of demoralization because the predictive value of each variable was assessed while all of the other significant variables were held constant.

Parallelling the strategy used with the Travis data, significant and trend-significant variables obtained in the F-change tests in Linden–Perth Amboy were entered simultaneously with levels of pre-incident psychological symptomatology. Perceived threat to physical health, having a pre-teenaged child and number of pre-incident psychological symptoms were significant in predicting levels of

.TABLE 6.8
F-change Statistics for Regression Predicting Demoralization After
Controlling for Pre-incident Psychological Symptoms

Variable	TRAVIS		LINDEN-PERTH AMBOY	
	F-change	p	F-change	p
1. Age	2.3	NS	0.1	NS
2. Sex	2.1	NS	0.0	NS
3. Marital Status	2.0	NS	1.4	NS
4. Education	3.1	TREND	0.0	NS
5. Years Lived in Area	3.6	TREND	2.7	TREND
6. Preteen-aged Child	6.4	SIG**	7.0	SIG**
7. Homeowner	0.2	NS	0.6	NS
8. Distance From Site	2.6	NS	0.9	NS
9. Date of Interview	5.3	SIG*	1.0	NS
10. Distrust	4.8	SIG*	2.2	NS
11. Attitudes Toward Move	19.4	SIG***	3.3	NS
12. Perceived Threat to Physical Health	33.7	SIG***	10.2	SIG***

NS; Not Significant
SIG; Statistically Significant
TREND; $p > .05$ and $< .10$
* $p < .05$
**$p < .01$
***$p < .001$

TABLE 6.9
Significant Predictors of Demoralization in Travis and Linden-Percy
Amboy

Variable	TRAVIS			LINDEN-PERTH AMBOY		
	Beta	t	p	Beta	t	p
Threat to Physical Health	.38	3.2	$< .01$	.21	2.2	$<.05$
Number Psychological Symptoms Before Incident	.38	3.7	$<.001$	.61	6.4	$<.001$
Attitudes Toward Moving	.26	2.3	$<.05$	Not Significant		
Having a Pre-teenaged Child	Not Significant			.21	2.4	$<.02$

demoralization in Linden–Perth Amboy. (See Table 6.9 for a summary of the multiple regression tests for both Travis and Linden–Perth Amboy.)

THE FACTORS THAT LEAD TO PSYCHOLOGICAL
DISTRESS AFTER A TOXIC EXPOSURE EVENT

In this chapter, we examined the relative value of a number of sociodemographic, situational, and attitudinal dimensions to predict post-incident psychological distress in two different communities that were affected by two independent hazardous chemical incidents. The major dependent study variable was level of demoralization, a measure of psychological distress (or well-being) that can be influenced by environmental stressors (Link & Dohrenwend, 1980), including a hazardous exposure incident (Dohrenwend et al., 1981). The two communities chosen for study differed across a wide range of survey characteristics, notably sample sex composition, number of pre-incident psychological symptoms, marital and homeowner status, perceived threat to physical health, and current levels of demoralization. In identifying predictors of demoralization following the chemical events we obtained similar results in our two communities, particularly with regard to the predictive role of perceived threat to physical health. Given the differences between the Travis and Linden–Perth Amboy communities and the dissimilarities between the two hazardous exposure incidents, the replication of our major findings is noteworthy. These results may be generalizable to toxic exposure-affected communities other than those examined in this chapter.

We consider the critical effect of perceived threat to physical health on levels of demoralization to be our most important finding. Although the specific items that compose the perceived threat to physical health scale tap views about being contaminated and whether one's health has deteriorated as a result of the incident, the measure appears to reflect concern regarding future health, awareness of environmental hazards, and attention to potential symptoms of exposure.

Mechanic (Mechanic, 1979, 1980) has studied the effect of symptom monitoring and introspection as it generally influences psychological health. This investigator holds the view that introspection is "a learned predisposition that is triggered by particular events" (Mechanic, 1983, p. 6). Similarly, we consider perceived threat to be an orientation that often is acquired through personal familiarity or direct experience with hazardous exposure events. Although high levels of perceived threat could be a result of poorer psychological health, we have ruled out this alternative explanation by controlling for psychological symptoms prior to the respective incidents at Travis and Linden–Perth Amboy.

In an attempt to address the question as to whether levels of demoralization found in our community samples were "high," we can compare our results with those obtained by the Behavioral Task Force of the President's Commission on the accident at Three Mile Island. These comparisons, however, should be interpreted in the context of the following considerations:

1. The demoralization measures used in all three studies, though very highly correlated, were not identical.
2. The President's Task Force used TMI-resident mental health clients as a criterion group to make estimates of the prevalence of demoralization in the community.
3. Sociodemographic compositions of the Travis and Linden–Perth Amboy samples may have been quite different from the TMI general population sample. Some of these dimensions, such as sex and economic status, are known to be correlated with demoralization and we were unable to control for these variables in making our comparisons.

Our data show that demoralization levels at Travis were about the same as levels found by the Behavioral Task Force at TMI in their general population study. These levels were considered high by the Task Force that estimated that a significant minority of general population residents had demoralization levels that were elevated based on levels found among mental health clients. The Task Force concluded that "the accident at TMI had substantial immediate psychological effects on the people living in the area" (Dohrenwend et al., 1981, p. 157). Using the TMI results as a benchmark, demoralization levels at Travis can be viewed as similarly elevated.

There was a trend for levels of demoralization in Linden–Perth Amboy to be lower compared to Travis. However, it should be kept in mind that unlike Travis and TMI, the majority of Linden–Perth Amboy subjects were male and that, in general, females have significantly higher levels of demoralization than do males (Link & Dohrenwend, 1980). The question then arises whether the differences in demoralization between Travis and Linden–Perth Amboy residents is associated with the sex compositions of the respective samples. Using t-tests, we ran within-sex demoralization levels across samples. These tests revealed that there was no significant difference in mean demoralization levels between Travis and Linden–Perth Amboy males ($t = 0.2$; $p > .8$) but that Travis females were significantly higher on demoralization than Linden–Perth Amboy females ($t = 2.0$; $p < .05$). Nevertheless, when we subsequently controlled for perceived threat to physical health using ANCOVA, the univariate sex effect disappeared. Thus, although levels of demoralization in Linden–Perth Amboy were lower than levels in Travis, this effect was completely mediated by differential levels of perceived threat to physical health. These results, once again demonstrate the importance of perceived threat to physical health in predicting demoralization. This result also raises the possibility that by lowering levels of threat, levels of demoralization could similarly be lowered.

Having a pre-teenaged child was a significant predictor of demoralization at Linden–Perth Amboy and Travis when we controlled for prior psychological symptoms. However, when levels of perceived threat were also held constant, this relationship remained for our Linden–Perth Amboy sample but disappeared among Travis residents. Among Linden–Perth Amboy residents there appear to be views regarding the increased vulnerability of young children that transcend perceived personal threat. Perhaps given the multiplicity of hazardous exposures at Travis and their more chronic nature, those who have chosen to remain there have discounted any special risks for their children.

At Travis, positive attitudes toward moving increased levels of demoralization even in the post hoc model that controlled for several other key dimensions. There was no such significant relationship between attitudes toward moving and demoralization at Linden–Perth Amboy. A measurement artifact may explain the difference in findings between samples. The attitudes toward move scale had acceptable alpha reliability coefficients in the studies conducted by the Presidents' Task Force at TMI (alpha = .79 to .81 in general population study) and in the Travis sample (alpha = .74). In contrast, among Linden–Perth Amboy residents, the scale had a substantially lower reliability coefficient (alpha = .41). We attribute this dissimilarity to the sex composition of the different studies. At Travis and TMI, about two thirds of the sample were female. In Linden–Perth Amboy, respondents were about 60% male. When we ran reliability coefficients for the move scale broken down by sex, the scale was more internally reliable for females (alpha = .68) than males (alpha = .28). This result demonstrates the importance of conducting reliability analyses on subgroups within a sample.

The single most powerful predictor of demoralization in both of our community samples was number of psychological symptoms before the incident. Certainly, those with psychological symptomatology that antedates the toxic exposure would be expected to be higher on post-incident demoralization. Future studies, which include samples of non-exposed controls, can better address the question as to whether this is a general or incident-related effect. If the effect is general, pre- and post-incident psychological symptom levels would covary for controls just as they would for experimental subjects. If the incident-related model held, exposure status and pre-incident psychological symptom levels would interact. Thus, those exposed and high on pre-incident psychological symptoms would be higher on post-incident distress relative to non-exposed subjects with similar high levels of pre-incident distress. Such an interaction would suggest that respondents with high levels of pre-existing psychological distress are particularly vulnerable to the stress of a toxic exposure event. Given either pattern, however, the use of retrospective self-reports to measure prior symptoms has obvious limitations, and researchers in this area must begin to develop design strategies that adequately control for this problem.

Pre-incident psychological symptoms correlated significantly with demoralization in Travis (Pearson's $r = .51$) and in Linden–Perth Amboy (Pearson's $r = .69$). The fact that the correlation was stronger in Linden–Perth Amboy compared to Travis may help explain why there were more variables in Travis that had significant F-change statistics. This statistic simply tells us whether additional variance in the dependent variable can be explained by other independent dimensions.

In contrast to sociodemographic factors that influenced levels of demoralization among community residents at TMI, in our sample these background factors were not significant predictors, once we controlled for prior psychological symptoms. Most likely, the higher levels of demoralization found among women at TMI compared to men is due to the lack of a measure of prior psychological symptoms.

There also is a major difference between the incidents at TMI and both Travis and Linden–Perth Amboy. At TMI, there were no immediate physical health effects attributed to exposure. In contrast, at Travis and Linden–Perth Amboy, numerous residents experienced physical symptoms resulting from exposure to chemicals in the air.

An important consideration in studying hazardous exposure incidents is the selection of control or comparison groups. In Linden–Perth Amboy, we employed a stratification procedure, choosing two "exposed" communities that differed in distance from the initial site of the leak. Nevertheless, there was no pattern of increased effects of the exposure in Linden relative to Perth Amboy. A likely explanation is that wind currents played a key role in disseminating the noxious fumes. Thus, in both Linden and Perth Amboy there were exposed and non-exposed areas of the community.

Because we did not have a control population for the Travis sample, we made use of a replication strategy, employing our samples from Linden–Perth Amboy. The comparability of our results despite major differences between the Travis and Linden–Perth Amboy communities reduces the chance that we are reporting false positive findings.

Little is known regarding causal pathways in the etiology of psychiatric distress and disorder following toxic exposure events. The identification of high-risk groups within the community may then be the single most useful strategy of dealing with the problem given our present state of knowledge. If an incident were to occur, crisis intervention could be focused on those members of the community most likely to be adversely affected by the event. Cohen and Ahearn (1980) advocate a needs-assessment survey of affected groups in postdisaster situations. However, this practice may delay help to those who need it most because field interviews and intensive history taking are time-consuming endeavors. It thus becomes important for planners and clinicians to identify psychologically vulnerable groups before an incident actually occurs.

If threat is a common pathway to distress, we may be able to intervene in this process. One plan for researchers is to attempt to generate predictors of threat that may be changed through effective community interventions. For example, if distrust in authorities were found to predict threat, community leaders could attempt to upgrade existing levels of trust even before potential incidents occurred. After a toxic exposure event, trust could be maintained by keeping channels of information related to the incident open to the public and by developing public policies to properly disseminate information. By making the nature of the threat as non-ambiguous as possible, levels of threat could be allayed.

According to Irving Janis (1962), normal reactions to disaster entail a balance between the need to maintain vigilance and a quest for reassurance that the danger is no longer threatening. Maladaptive patterns are likely to ensue when the perception of threat is so strong that it prevents assimilation of reassuring information. Affected community residents have to be provided with adequate sources of honest and reassuring information, so that a balance can develop between vigilance and security. In future studies, we recommend that more attention be given to factors underlying perceived threat. In addition, aspects explaining perceived confidence (that a danger is not threatening) could be addressed as well.

Toxic exposure incidents cannot be viewed solely in the context of their possible adverse physical health sequelae. Both short- and long-term mental health effects need to be systematically assessed using reliable and valid measures along with scientifically appropriate research designs.

Now and in the future, toxic exposure events are likely to be an enduring fact of life in industrialized societies. We must begin to develop methods of identifying those members of the community who may be psychologically injured by these incidents and attempt to reduce these mental health effects with effective interventions.

ACKNOWLEDGMENTS

The two community studies were non-funded research projects paid for by the authors. However, the computer needs of the studies were supported by a grant from the National Institute of Mental Health (MH-30906-06) to the New York State Psychiatric Institute. We are grateful to Matthew Tarran for helping with the field work and coding.

REFERENCES

Bromet, E., Schulberg, H. C., & Dunn, L. (1982). Reactions of psychiatric patients to the Three Mile Island nuclear accident. *Archives of General Psychiatry, 39,* 725–730.
Cohen, R., & Ahearn, F. (1980). *Handbook for mental health care of disaster victims.* Baltimore: Johns Hopkins University Press.
Dohrenwend, B. P., Dohrenwend, B. S., Warheit, G. J., Bartlett, G. S., Goldsteen, R. L., Goldsteen, K., & Martin, J. L. (1981). Stress in the community: A report to the President's Commision on the Accident at Three Mile Island. *Annals of the New York Academy of Sciences, 365,* 159–174.

Dohrenwend, B. P., Shrout, P. E., Egri, G., & Mendelsohn, F. S. (1980). Specific psychological distress and other dimensions of psychopathology. *Archives of General Psychiatry, 37,* 1229–1236.

Epstein, S. S., Brown, L. O., & Pope, C. (1982). *Hazardous wastes in America.* San Francisco: Sierra Club Books.

Fleming, R., Baum, A., Gisriel, M. M., & Gatchel, R. J. (1982). Mediating influences of social support on stress at Three Mile Island. *Journal of Human Stress, 8,* 14–22.

Frank, J. (1973). *Persuasion and healing.* Baltimore: Johns Hopkins University Press. (Originally published 1961)

Goldsteen, R. L. (1983). *The Three Mile Island accident: A case study of life event appraisal.* Unpublished doctoral dissertation, Columbia University School of Public Health.

Janis, I. (1962). The psychological effects of warning. In G. Baker & D. Chapman (Eds.), *Man and society in disaster* (pp. 55–92). New York: Basic Books.

Kasl, S. V., Chisholm, R. F., & Eskenazi, B. (1981). The impact of the accident at the Three Mile Island on the behavior and well-being of nuclear workers: Part II: Job tension, psychophysiological symptoms, and indices of distress. *American Journal of Public Health, 71,* 484–495.

Lebovits, A. H. (1984, August). *Case of asbestos-exposed workers: A psychological evaluation.* Paper presented at the 92nd Annual Convention of the American Psychological Association, Toronto, Ontario.

Link, B. G., & Dohrenwend, B. P. (1980). Formulation of hypotheses about the true prevalence of demoralization in the United States. In B. P. Dohrenwend, B. S. Dohrenwend, M. Schwartz-Gould, B. Link, R. Neugebauer, & R. Wunsch-Hitzig (Eds.), *Mental illness in the United States: Epidemiological estimates* (pp.   ). New York: Praeger.

McFadden, R. D. (1984, October, 7). Scores are felled by cloud of fumes. *New York Times,* p. 1, 40.

Markowitz, J. S., & Gutterman, E. M. (1984, August). *Predictors of demoralization following a toxic chemical waste incident.* Paper presented at the 92nd Annual Convention of the American Psychological Association, Toronto, Ontario.

Markowitz, J. S., Gutterman, E. M., & Link, B. G. (in press). Physical and psychological effects among seamen following a malathion pesticide incident. *The Journal of Occupational Medicine.*

Mechanic, D. (1979). Development of psychological distress among young adults. *Archives of General Psychiatry, 36,* 1233–1239.

Mechanic, D. (1980). Experience and reporting of common physical complaints. *Journal of Health and Social Behavior, 21,* 146–155.

Mechanic, D. (1983). Adolescent health and illness behavior: Review of the literature and a new hypothesis for the study of stress. *Journal of Human Stress, 9,* 4–13.

Pfeiffer, A. (1980, April 25). Fourteen dangerous chemicals among those leaked into Kill. *Staten Island Advance,* p. 97.

Pfeiffer, M. B., Patrick, D., & Cross, S. (1980, April 22). Islanders have to hold their breath as Jersey chemicals go up in flames. *Staten Island Advance,* p. 96.

United States Senate. Report of the Subcommittee on the "Potential Health Effects of Toxic Chemical Dumps" of the DHEW Committee to Coordinate Environmental and Related Problems, in *Health effects of toxic pollution,* Report to the U.S. Senate Committee on Environmental and Public Works, (Ser. 96–15), August, 1980, Washington, D.C.

7 Threats to People and What They Value: Residents' Perceptions of the Hazards of Love Canal

Adeline G. Levine
Russell A. Stone
State University of New York at Buffalo

The social consequences of hazardous waste disposal at Love Canal provide a landmark example of a new sort of human problem emerging from development in industrial societies. Although Love Canal has been called a *disaster*, both by the media and in the two major social science studies about the events at that site (Fowlkes & Miller, 1982; Levine, 1982), in fact it is a new sort of disaster, or more generally, a new kind of social crisis event (Quarentelli & Dynes, 1977). These events do generate widespread misfortunes, thus satisfying a dictionary definition of disaster. However, because they are caused by people rather than by the forces of nature alone, they are seen as preventable. Such events are characterized by controversy and conflicting interpretations of what happened, and to whom, what policies should be followed in coping with the situation, and who should be responsible for carrying out those policies. Rather than starting like a hurricane or volcanic eruption, with a tremendous, visible impact occurring at a confined moment in time, they often are slow in onset, taking years, or decades to develop. In cases like Love Canal, an existing environmental condition, which may or may not be generally known about, is officially declared to be hazardous by agencies responsible for the health and welfare of the public. The physical consequences may take years to develop or become evident, may result in chronic conditions, or may be caused by other factors. As a result of the slow onset, and the shortage of technical information about the effects of low-level, long-term exposure to hazardous substances, it is difficult to establish direct causal relationships between the environmental condition and damage to people or property. However, like other recent, well-known problems, such as Three Mile Island, Agent Orange, and Times Beach, the Love Canal situation falls under Barton's (1969) definition placing disasters in a more general category of:

> collective stress situations . . . (where) . . . many members fail to receive expected conditions of life from the system . . . includ(ing) the safety of the physical environment, protection from attack, provision of food, shelter, and income, and guidance and information necessary to carry on normal activities. (p. 38)

The Love Canal residents whose responses are reported here were in the midst of a collective stress situation at the time they were interviewed. Thus, this research differs from studies of disasters with swift impact that are necessarily done after the fact. Because of the nature of the Love Canal events, we were able to interview and observe people while the crisis was in process. Neighborhood residents were slowly growing aware of how the events were affecting their lives. They were trying to comprehend that physical conditions in their immediate environment, which were previously unknown or simply accepted as harmless, might be hazards, "threats to people and what they value" (Kates & Kasperson, 1983, p. 7027). The materials were gathered as part of a larger study at Love Canal (Levine, 1982) of people encountering newly discovered hazards, how they perceived their problems, and how they were managing them. Such studies help to fill the gap in human risk analysis noted by Short (1984) who pointed out that studies of risk "have virtually ignored how people in fact live with risks and how living with risks affects them" (p. 712).

This chapter focuses on Love Canal residents' perceptions of the problems they were facing, and the changes in their own lives from the time the crisis began for them. Historically, the events began in 1953, when the Hooker Chemical Company sold the Love Canal property to the Niagara Falls, New York school board for a token $1. The canal itself was an abandoned, covered-over short waterway, which the company had filled with 22,000 tons of residue from the manufacture of over 200 chemical compounds. In 1955 an elementary school built atop the canal opened its doors to 400 children, and a residential neighborhood of small homes grew up around it. Eventually 1,000 families, including about 250 tenants of federally subsidized housing, lived within the 10-block area surrounding the canal. Those streets became known as the Love Canal neighborhood.

By the mid 1970s it was apparent that chemicals were leaking from the disposal site. After 2 years of quiet, preliminary studies by governmental agencies, the residents suddenly were alerted to the possibility that they were living in an environmentally unsound area. In the late spring of 1978, the New York State Department of Health announced that it was about to conduct epidemiological studies of health conditions in the neighborhood, that the Federal Environmental Protection Agency was studying chemicals in air samples from the basements of some homes, and that the state Department of Environmental Conservation was studying materials taken from basement sump pumps and from storm sewers adjacent to the area. Within weeks, it was determined that vapors from benzene, lindane, trichloroethylene, and several other highly toxic chemicals were present in the air in some homes, and that women living in homes adjacent to the old disposal site property had elevated miscarriage rates.

On August 2, the State Commissioner of Health issued an official order stating that the Love Canal was a "great and imminent peril to the health of the general public residing at or near the site" (Whalen, 1978, p. 27). The Commissioner issued orders to close the elementary school temporarily, to implement a remedial engineering plan to stop the migration of toxic substances, and to conduct further environmental and health studies. He also recommended that pregnant women, and children under age 2 immediately move from the area adjacent to the canal, and that all residents should avoid using their basements and eating vegetables from their gardens.

Within hours the citizens formed an action group. Within days the President of the United States declared that an emergency existed at Love Canal, and the Governor of New York personally told assembled residents that the 239 families living nearest to the disposal site, in the "inner ring," could sell their homes to the state and receive assistance in moving to new homes.

In order to carry out that offer, to implement the remedial construction project, and to launch a massive epidemiological study of the health of the Love Canal neighborhood residents, the Governor established a 10-agency task force that set up headquarters at Love Canal. People began to move out from the inner-ring homes. The Department of Health distributed hundreds of health question-naires and took thousands of blood samples. In October, dozens of workers and huge construction machines invaded the neighborhood, beginning the massive construction project to recap and partially drain the old buried canal.

All these events earned ongoing worldwide publicity. The active citizens' group concentrated first on the inner-ring residents' expressed desire to be moved out of their homes. Then it turned to the fate of the outer-ring residents, people living in the remainder of the Love Canal neighborhood, with a closed school and hundreds of rapidly emptying homes in the center. A 2-year-long struggle ensued between the organized residents concerned about the threat to their health and their suddenly worthless homes, and local, state, and federal agencies that seemed bent on minimizing the extent and seriousness of the migration of chem-icals and health problems as the agencies came to realize the staggering immediate costs and the implications of setting a precedent of governmental responsibility for future Love Canals (Levine, 1982).

In late August, the state health department distributed a report dramatically titled *Love Canal: Public Health Time Bomb* (NYDOH, 1978). People were moving out at state expense, a giant construction project was under way to control the migration of toxic wastes, and a massive health survey was in process. In November, a state health official announced that liver function tests indicated abnormalities among some young boys and men. In December, the state health department announced at a community meeting, with prepared statements for the press, that dioxin had been located on some properties, and in storm sewers and creeks in the Love Canal neighborhood. Despite this evidence that something serious had happened, residents were subject to conflicting messages from the authorities. Although the area was deemed a health hazard in general, people were reassured in public and private meetings that there was "no evidence" of harm to specific individuals.

By the early autumn of 1978, when the first wave of interviews discussed in this chapter was conducted, the collective stress situation for the community had shifted from the unexpected crisis of learning about the hazards, to a long period of psychological "attrition," resulting from "cumulative exposure to a threat." This is a particularly demoralizing period, because the "worst kind of threat . . . (is) . . . the generalized dread of the unknown" (Lang & Lang, 1964, p. 71). At Love Canal, the situation was exacerbated by the conflicting messages the residents received constantly.

Although full results of the health department's massive health survey were never made public, New York's new Commissioner of Health announced on February 8, 1979, that upon examining the results of the survey, he had decided to extend the previous health order. The Commissioner disclosed that there were excessive numbers of miscarriages and low birth weight babies among women living in outer-ring homes, located on covered-over underground drainage gullies and ponds. In the new health order, the Commissioner's recommendation to the Governor was limited: Pregnant women and children under age 2 (and their immediate families) should be moved out at state expense, until the youngest child was 2 years old. This recommendation was opposed by the organized residents who argued that damage to fetuses implied potential health problems not just for fetuses but for everyone. Love Canal residents were increasingly mistrustful of the official descriptions of their health conditions. They felt forced to rely on their own resources. The second wave of interviews discussed here began shortly after the supplemental health order was issued in February 1979, while the residents were assimilating the latest conflicting information.

The findings described in this chapter focus on some of the respondents' perceptions of how the Love Canal situation was affecting them and their families in the autumn of 1978 and the spring of 1979. Comparative findings from a study started in the spring of 1980 (Fowlkes & Miller, 1982) are mentioned where relevant. We begin by describing what respondents viewed as the major problems for their families, and then go into detail on three aspects of the problems: health, homes and property, and effects on personal outlook and interpersonal relations.

We discuss differences in perceptions, and show correlations with nine social background variables: (a) age of respondent or head of household, (b) youngest

child's age, (c) number of children, (d) family income, (e) education of respondent or head of household, (f) length of residence at Love Canal, (g) home
ownership versus renting, (h) location of residence, inner ring (the 239 single-
family homes closest to the canal) versus outer ring (the homes and apartments
in the remainder of the neighborhood), and (i) the time when the respondent
first became aware of the Love Canal situation (before 1975, 1975–1977, or not
until 1978).

METHODS

The interviews were conducted by Levine and a team of five sociology graduate
students. During the weeks between late October and early December 1978, 102
members of 58 Love Canal families participated in the first wave of interviews,
which lasted from 1 to 4 hours. The interviews were conducted in the respondents'
homes. An open-ended, semi-structured, purpose-designed interview form was
used, which allowed all interviewees to respond to a fixed set of questions at
length. Respondents were selected to represent a wide range of points of view
and social characteristics, but the sampling design was not random or formally
systematic. We have termed it an *informal snowball* design. Beginning with
families of the emerging leaders in the community, early respondents were asked
to suggest others who did not feel as they did. Other potential participants were
selected from membership lists of citizens' organizations. Care was taken to
include in the sample people of different ages, residence locations, and ownership
status (homeowners and renters). Families who had moved away as well as those
who remained were included. Comparisons of the social background characteristics of our sample with the more systematically drawn sample of Fowlkes and
Miller (1982) indicate that this sample was well representative of the population
of about 1,000 families then living in the Love Canal area.

In the second wave, 64 members of 39 of the families were reinterviewed in
their homes, in the spring of 1979. By this time, more of the families had moved,
and additional evidence of toxic chemicals and health problems had been pub-
licized, with attendant controversy over the meaning and accuracy of the findings.
Not all the families originally interviewed could be contacted because some had
moved away, some refused a second interview, and research resources (time
and funds) were limited. However, the second wave retained representativeness
on the distribution of background characteristics of respondents.

The unit of analysis in this study is the family, rather than the individual
respondent. In the first wave, 36 husbands and wives were interviewed together,
as were 25 in the second wave. The remainder of the interviews were with one,
or occasionally three adults in the family, including single-parent or single-head
households, plus families where only one spouse was present at the time of

interview (the women in all cases but one). The coder, working from transcripts of the interviews, was instructed to identify the opinions and perceptions upon which the respondents agreed. Where relevant, demographic characteristics of the "head of household" are used to characterize the family (e.g., age, education).

FINDINGS

Perception of the Problem

First, to what extent did Love Canal residents view the entire situation as problematic for themselves? In the fall, the respondents were asked: "Is the Love Canal situation a problem for you? . . .for your children?" Only 12% claimed it was not a problem, with the overwhelming majority naming more than one. This question, as most of those reported here, elicited lengthy answers that were subsequently coded for content. Table 7.1 indicates the major categories of problems mentioned by respondents. Multiple responses were coded so the percentages total more than 100.

The problems described most frequently were physical health and worry/anxiety. This was a time when people felt they had lost control of major aspects of their lives. Many expressed uncertainty as they reflected upon their lives, wondering whether their families' illnesses were related to chemical exposure, whether they or their children would be prey to unknown chemical-related illnesses in the future, whether they should move (either with governmenal help, or on their own), and whether they could shoulder the financial burden for years to come. Some feared that they might unwittingly move near another hidden disposal site. Some felt the authorities were not giving them helpful, credible guidance for decision-making, and they felt pressured to make decisions quickly, based on inadequate information (see Janis & Mann, 1976).

These working-class people were faced with real threats to the economic security represented by their homes. Half the respondents mentioned property damage or drop in property value as a problem, with 12% mentioning other financial loss. From the beginning of the crisis, unsympathetic critics accused Love Canal residents of being concerned chiefly with property values, some

TABLE 7.1
Perceptions of Problems

Percent mentioning:

Physical Health	70%
Worry, Anxiety	68%
Property Damage, Loss	51%
Other Financial Loss	12%
No Problem	12%

going so far as to charge that the residents had seized upon a golden opportunity to "make a bundle off the government." However, our interviews (and observations) showed that physical and emotional factors were of primary concern to the residents. The coder was asked to judge the problem considered most serious by respondents, usually the one mentioned first. In no instance was property or financial loss perceived as the foremost problem. Although the worry and anxiety may well have been related to the loss of homes, with the attendant emotional and financial implications those losses would entail, no interviewer or observer working on this study had the impression that any Love Canal resident welcomed the possibility of a coerced move.

Given the large proportions of residents indicating physical health and mental well-being as problems, there is little to be learned from systematic examination of all the background variables. However, we found that a larger proportion of inner-ring respondents mentioned "physical health" as a problem (82% vs. 65%), while outer-ring respondents were slightly more likely to mention "worry, anxiety" (70% vs. 65%). The inner-ring residents had been closer to the sources of contamination, whereas the outer-ring residents were grappling with the problems of being left behind in a semi-abandoned neighborhood.

Of the inner-ring people, 71% mentioned "property damage or loss" as problems as against 43% of outer-ring respondents at the early stage in the crisis. Economic loss was more of a concern among families without young children, those with higher incomes, and those who had lived in the area longer. Naturally, no renters expressed concern over property damage or economic loss. We have reported elsewhere that families with "new roots" in the community (homeowners, 6–10 years in the community, with young children, moderate incomes, and better educations) were more likely to be activists (Stone & Levine, 1985). They also were the ones most concerned with physical and emotional health rather than property damage.

Perceptions of Health

Two questions in wave 1, and a follow-up in the wave 2 interviews focused specifically on respondents' perceptions of their health status and related behavior. "Is your health better or worse, or about the same as it was a few years ago? . . . (or) since you moved here?" and "Are you visiting doctors more?" were asked in wave 1. Answers from 54 families were coded. In wave 2 the question asked was: "Has anything happened to your health and to your family's health since the last time we spoke?" In response to this question, the respondents provided answers for every wave 2 family member, 103 people. Distributions of responses and correlations with the nine background variables are listed in Table 7.2.

In fall 1978, almost two thirds of the respondents reported that their health worsened during residence at Love Canal (HEALTH). Not everyone reporting

TABLE 7.2
Perceptions of Health
Response Distributions and Correlations—Gamma (*)—with Background Variables

	(1) AGE + young	(2) KIDIAGE + young	(3) KID-NMBR + more	(4) INCOME + higher	(5) EDU-CATN + more	(6) RESYRS + longer	(7) OWN-RENT + owner	(8) RING + inner	(9) AWARLC + earlier
HEALTH (n = 54) worse—63% same—33% better—4%	.35 (*)	.44	.19	.12	.10	-.10	.34	.33	.05
VISIT DOC (n = 41) more—46% same—49% less—5%	.13	.21	.19	.19	-.21	.20	.12	.04	-.05
HLTHCHG2 (n = 103) worse—26% same—35% better—39%	.07	-.29	.06	.07	-.11	.02	-.50	-.05	-.04

(1) AGE is age of respondent, or head of household.
(2) KIDIAGE is age of youngest child in the family.
(3) KIDNMBR is the total number of children in the family.
(4) INCOME is total family income.
(5) EDUCATN is number of years of schooling of respondent, or head of household.
(6) RESYRS is number of years of residence in the Love Canal neighborhood.
(7) OWNRENT is whether respondents are homeowners or renters.
(8) RING is location of Love Canal residence, inner ring (nearest the canal) or outer ring.
(9) AWARLC is when respondent was first aware of possible danger from chemicals buried at Love Canal.
(*) The correlation coefficients reported in Tables 7.2 to 7.5 are gamma, a coefficient used with nominal and ordinal data. It is the most appropriate measure for the data in this study, as in all cases the dependent variables are ordinal. The few continuous independent variables (age, income) were grouped into ordinal categories. Significance levels are not usually calculated for gamma, and are not relevant here because the sample was not a random one drawn from a large, normal population.

worse health, however, sought more medical attention than they had in the past (VISITDOC). There are at least three possible explanations for this seeming discrepancy. First, we did not ask for precise information about the number of visits the families customarily made to doctors; no increase in visits might well have been occurring against a pattern of frequent visits that gave people the chance to tell their doctors about increased problems. Second, some of the health problems reported were of a chronic, low level nature, such as excess fatigue, mild skin rashes, headaches, and allergies, the sorts of problems that people try to care for themselves. Third, a theme expressed by many respondents was that they had told their doctor about their problems, the doctor had been unable to diagnose them, so they had stopped trying to get help from their physicians. This is consistent with Fowlkes and Miller's (1982, p. 101) finding that the health complaints at Love Canal frequently did not fit easily diagnosable patterns.

The questions about health change asked in wave 2 elicited data on each family member (103 people in the 39 families reinterviewed). The distribution of responses (HLTHCHG2) shows more people reporting improved health than deteriorating health. The reversal of the trend from the first health question in wave 1 is quite striking. From the individual interviews, we know that much of this reversal can be attributed to families who moved from Love Canal in late 1978 or early 1979. Also, an informal telephone survey conducted by the organized residents in February 1979 found that 67 of 101 former inner-ring residents reported improved health 32 reported no change, and only 2 felt worse, since their move a few weeks or months before (Levine, 1982, p. 100).

Turning to the correlations with background variables in Table 7.2, the coefficients reported—gamma—show the strength and direction of relationship between the perceptions of health, and each of the background variables. A positive coefficient indicates that the "+" characteristic (young, in the case of AGE) is more likely to be reported among the first category listed for each perception. For example, the .35 under AGE indicates that younger respondents were more likely to report worse health. As a rule of thumb, in Tables 7.2 to 7.5 we take a correlation of .20 or greater as indicating a relationship worthy of note.

Thus, in the fall of 1978, shortly after the crisis began, younger heads of households, families with younger children, and homeowners, particularly those in the inner ring were more likely to report worse health in their families over the last few years. These were the families most sensitive to the crisis both on the basis of their residence location closest to the old canal, and because of the concern over vulnerability of young children. The pattern weakens when we examine the second health perception (VISITDOC). There is no correlation stronger than .21. Families with less education reported visiting doctors more, as shown by the negative coefficient. Families with young children visited doctors more than those with older children, which is likely in any population.

A few months later, in wave 2, there were few relationships between reported health change and background variables, except that families with young children were less likely to report worsened health, as were homeowners. All inner-ring people were homeowners, so more of them were likely to have moved from the Love Canal neighborhood. These findings about improved health upon leaving the Love Canal dwellings corroborate the findings of the telephone survey of 101 former inner-ring families just mentioned. To our knowledge, these are the only available longitudinal data about the general health of the residents of Love Canal. No follow-up research has been done by the responsible health authorities.

Perceptions of Home and Property

Not only did Love Canal residents have to evaluate their own health status with ambiguous data and standards, but also, those evaluations were intertwined with decisions about leaving their homes and moving from the neighborhood in which they had built their lives. This was a particularly anguishing decision for home-owners. Their houses were their principal financial assets. Some men and women had worked several jobs simply to amass the down payments. Many made extensive improvements to their properties, and now faced the prospect of leaving their homes to possible vandalism (which did occur after houses were left empty), then gambling on eventual sale or rehabilitation. If they were inner-ring residents they faced the prospect of selling to the government at a price that would not permit them to replace with equivalent housing unless they invested additional money, or could meet much higher monthly payments.

From our interviews and observations we were impressed with the emotional impact of the loss of home. For some it was the physical loss of the house itself. For others it was loss of the feeling of safety and security that the term *home* connotes. Most people described their homes as the center of their daily lives, their recreation, their work, and as a source of autonomy and status. Many people seemed almost to be in a period of mourning for their homes when we first interviewed them. Some people who had moved returned frequently, just to look at their former houses. Home was described as the heart of the family's life with poignant expressions of longing for the lost object and for the lost feelings about the home. "I keep thinking I'm going to go back, that I'll wake up in my own little home," said a woman who had moved from the inner ring. "I had big plans for my place, but now it's a dead box and I'm not going to keep trying to fix it up," said an outer-ring man, disappointed about the house he had helped to build, and now wanted "to get rid of, any way possible."

Perceptions of whether their homes were safe and whether they wanted to move were undoubtedly affected by complex combinations of health and property considerations. These attitudes were elicited in both waves of interviews. In the autumn, people were asked: "Do you feel your home is safe?"; "Do you want to move?"; "What kind of home are you looking for, and where?" The following spring, they responded to: "Do you feel safe in your home now?" and "Do you wish, at the present time, to leave the Love Canal area?" The distribution of responses and correlations with background variables, where relevant, are shown in Table 7.3.

The marginal distributions on perceptions of home safety shifted sharply between waves 1 and 2. In the autumn of 1978 the majority (67%) felt their homes were not safe (HOMESAF1), but by spring 1979, only 38% of respondents still felt their homes weren't safe (HOMESAF2). This can be explained by the relocation that had taken place between the waves of interviews. In wave 1, younger families with young children, homeowners, in the inner ring were most likely to feel their homes were unsafe, and these were the categories of people most likely to have moved during the first round of evacuations and home purchases. Thus, those remaining in the inner ring by the time of wave 2 were more likely the ones who felt their homes were safe. Also, those who had moved were likely to feel that their new homes were safe.

Another strong explanatory variable for lack of feelings of safety in the first wave of interviews was income. Higher income, together with relatively younger age for adults and children, and slightly smaller (possibly incomplete) families characterize the families with "new roots," which we have seen to be the families most affected by the crisis (see Stone & Levine, 1985) and most likely to view their homes as unsafe.

Those who said that they knew about the chemical contamination earlier, before 1975 (AWARLC) were less likely to consider their homes unsafe. This relationship fades by wave 2, but in the early months of the crisis it appears that those who had become aware of the existence of dangerous chemicals most recently were most likely to consider them hazardous. Similarly those living in the area longer (RESYRS) and those who were older (AGE) were less likely to consider their homes unsafe. In the absence of a declared crisis, the chemical contamination may not have been perceived as dangerous, even though it was known to be present. Perhaps the longer one lives with potential dangers, the more likely one will become reconciled to the condition, if the effects are not acute in nature.

By the time of the second wave of interviews, the correlations show a shift from inner ring to outer ring residents (RING), and from homeowners to more renters (OWNRENT) feeling their homes were unsafe. The rental units were in the outer ring, so the two findings are related. The elevated levels of concern among families with young children persists in wave 2 (KID1AGE). The shifts in focus of concern indicate that over the initial months of the crisis more and more families came to perceive the situation as hazardous. Fowlkes and Miller (1982) reported that by the summer of 1980, only 9 (14%) of the 63 families they interviewed believed that the chemical contamination from Love Canal was limited in scope and provided minimal risk to health.

Questions on desire to move (MOVENOW1 and MOVENOW2) were asked only of those who had not been relocated at the time of interview. The proportion desiring to move from Love Canal rose from 76% in wave 1 to 86% in wave 2. In addition to coming to believe that their homes were unsafe, those remaining

TABLE 7.3
Perceptions of Home and Property
Response Distributions and Correlations (Gamma) With Background Variables

		(1) AGE + young	(2) KID-IAGE + young	(3) KID-NMBR + more	(4) INCOME + higher	(5) EDU-CATN + more	(6) RES-YRS + longer	(7) OWN-RENT + owner	(8) RING + inner	(9) AWARLC + earlier
HOMESAF1 (n = 42)	no —67% yes —33%	.73	.77	-.23	.53	.16	-.28	.72	.48	-.32
HOMESAF2 (n = 34)	no —38% yes —62%	-.01	.21	.13	-.09	-.05	-.10	-.24	-.51	-.19
MOVENOW1 (n = 50)	move far —56% move near—20% stay L C—24%	.56	.62	-.09	.56	.35	-.17	.73	.26	-.08
MOVENOW2 (n = 24)	move —86% stay —14%									
FINCHG2	worse —44%	.57	.18	0	-.05	.48	-.05	.44	.40	.02

(1) AGE is age of respondent, or head of household.
(2) KID1AGE is age of youngest child in the family.
(3) KIDNMBR is total number of children in the family.
(4) INCOME is total family income.
(5) EDUCATN is number of years of schooling of respondent, or head of household.
(6) RESYRS is number of years of residence in the Love Canal neighborhood.
(7) OWNRENT is whether respondents are homeowners or renters.
(8) RING is location of Love Canal residence, inner ring (nearest the canal) or outer ring.
(9) AWARLC is when respondent was first aware of possible danger from chemicals buried at Love Canal.

saw their community changing with neighbors moving away, empty houses boarded up, remedial construction work ongoing at the canal site, and the empty elementary school closed in the fall of 1978. Although the format of the wave 2 question did not preserve the distinction between desire to move far away, or stay nearby, the wave 1 data show that most of those desiring to move wanted to put distance between themselves and their former homes, reflecting a strong perception of the physical and emotional hazard of the area. Desire did not match with reality in many cases; our informal follow-up of the situation indicates that the majority of those relocated reestablished new homes close by. Jobs, kinship, friends, and long established community ties probably explain the inability of most people to simply "pull up stakes" and move far away, even after a large-scale crisis.

Very few renters considered their homes unsafe, and few wanted to move, possibly because the rental housing project in which most lived was considered the nicest subsidized housing project in all of Niagara Falls. The quality of housing and life could not be replaced for those on welfare or limited budgets. Perceived ability to make a change may have played a part as well. "What difference does it make if I think Love Canal is dangerous? Nothin' I can do about it no matter what I think," was one elderly, poor woman's comment.

The pattern of "new roots" background characteristics correlated strongly with desire to move in the autumn of 1978. The correlations with income may indicate more ability to afford a move, whereas the correlation with education may indicate that people with better educations realistically view themselves as more mobile in terms of job skills, and they may be more open to change.

Correlations are not reported for the desire to move in the spring of 1979 (MOVENOW2) because the large majority indicating a desire to move (86%) wiped out most possible variation. However, we found that all the younger respondents, all those with young children, all those with higher incomes, and all remaining inner-ring residents expressed a desire to move.

Finally on Table 7.3, a question asked in wave 2 focused on changes in financial situation: "Are there any changes in your financial picture since the last time we spoke?" (FINCHG2). The question was intended to focus on differences between the time of the first and second interviews. Almost half the families reported deterioration in economic well-being, although this might have been partly a reflection of depressed economic conditions in the region, not specifically the effect of the Love Canal crisis. However, we do know that families who moved, temporarily or permanently, incurred additional costs beyond the purchase price and supplemental funds provided by the government. Many home-owners perceived a drop in their property values, even if they had no intention of selling or leaving their dwellings. Thus, the impact of the real estate market was a clear financial hazard in that situation. This explains the strong correlation (.44) between home ownership (OWNRENT) and perception of financial change for the worse. It was particularly marked among inner-ring residents (all home-owners) and among younger, better-educated respondents. For most other respondents, financial well-being was stable with but one family reporting improvement.

Another item in the wave 2 interview asked: "Has anything happened to your property or home since we last talked?" Half those responding (15/30) reported some change. The state had purchased six of the homes, two had been damaged by chemicals and five by vandalism and fire. Surprisingly, two families reported repairing and improving their Love Canal homes between fall 1978 and spring 1979. The fact that some people still chose to invest in their property shows the ambiguity and differing interpretations of the situation perceived by various residents during the period of turmoil and conflicting factual evidence early in the crisis.

Outlook and Interpersonal Relations

Toward the end of the first interview there were two questions designed to probe people's sense of the impact the situation was having upon them personally. The first question was: "In a nutshell, do you feel optimistic or pessimistic?" This question was answered rather succinctly, and as shown on Table 7.4 (OPTI-MISM), the responses were almost evenly divided between optimism and pes-simism about the overall situation. Homeowners were particularly likely to feel optimistic. The longer ago respondents were aware of the potential problem (AWARLC) the more likely they were optimistic. Given the low correlation with age, and the slight correlation with years of residence (RESYRS), we can speculate that these were the older residents who had been living with the situation for years before it was defined as hazardous. They may have felt "psychically prepared" for the crisis, may not have defined it as serious at that point, thought that the remedial measures would be successful, or perhaps were keeping their spirits up in the face of a situation they could not influence. The only other strong correlation was between optimism and higher income. Those people felt more at ease with the situation because if necessary, they could afford to change their living situation with their own means.

The second question and related probes were: "What about you? How have you changed? What's happened to you personally? Are there things you found you can do that you never thought you could, or things you can't do that you may have thought about (doing) before?" These questions about personal changes elicited long responses, from which the coder interpreted the spirit as better, neutral, or worse (PERSCHNG). Beyond the summary statistics in Table 7.4, however, the answers given at length provided a wealth of detail.

A large majority of respondents experienced negative personal changes early in the crisis. People repeatedly expressed the idea that they had lost control over major aspects of their lives. That loss of control stemmed from someone else's decisions, long ago, to bury chemicals and to build homes near that site, not from any decisions the Love Canal people had made themselves. Now control

TABLE 7.4
Personal Outlook
Response Distributions and Correlations (Gamma) with Background Variables

	(1) AGE + young	(2) KID1AGE + young	(3) KIDNMBR + more	(4) INCOME + higher	(5) EDUCATN + more	(6) RESYRS + longer	(7) OWNRENT + owner	(8) RING + inner	(9) AWARLC + earlier
OPTIMISM (n = 44) optimistic—46% neutral—7% pessimistic—48%	.05	−.16	0	.35	−.17	.23	.74	.23	.43
PERSCHNG (n = 47) better—23% neutral—15% worse—62%	.25	.18	.12	.14	.12	−.16	.25	−.14	.35
LONGTERM2 (n = 32) yes—78% no—22%	.50	.53	−.50	−.23	.56	−.53	.23	.40	.17

(1) AGE is age of respondent, or head of household.
(2) KID1AGE is age of youngest child in the family.
(3) KIDNMBR is the total number of children in the family.
(4) INCOME is total family income.
(5) EDUCATN is number of years of schooling of respondent, or head of household.
(6) RESYRS is number of years of residence in the Love Canal neighborhood.
(7) OWNRENT is whether respondents are homeowners or renters.
(8) RING is location of Love Canal residence, inner ring (nearest the canal) or outer ring.
(9) AWARLC is when respondent was first aware of possible danger from chemicals buried at Love Canal.

rested in the decisions being made by political figures far away in the state capitol, and in Washington, DC.

This sense of loss of control affected their feelings about how they were performing their family roles. Men in particular spoke of feelings of helplessness, of not being able to provide a safe home for their families. Parents felt frightened about their children's health in the future. Many spoke of how the situation had interrupted the basic, normal routines. Daily conversations, patterns of child care, cleaning, shopping, laundry, family gatherings, birthday parties for children, other special events, and customary recreational activities had all changed. Many were weary from the demands of the experience, and felt they were not doing their work properly. A number of people suddenly experienced an unexpectedly broad range of powerful emotions. Some of the activists, for example, were surprised at their own behavior when they took vocal, belligerent public stances and did things they had never dreamed of doing in the past (see Gibbs, 1982).

On the other hand, some of the changes were perceived as positive ones. The younger adult respondents and homeowners were slightly more likely to perceive positive personal change. Some of them worked in the organized citizens' group, and they felt they were acting to change their life conditions more than usual. There was a whole new awareness of the issue of control of one's life, and of the importance of participating in the workings of government and the community by attending public hearings, by voting, and by taking roles in various organizations. One women told us she had learned to drive because she didn't want to feel dependent on others to get her out of the area if need be. Another told us that she had gained self-respect from her newfound ability to manage her home and family care in the midst of a crisis. Others spoke of how much they had learned about chemicals, about how government operates, and about how to create a citizens' organization. Still others were proud that they were managing to survive the stress. One couple had a profound religious experience during this period and were certain that God would help them to take care of the problems they faced.

In the second wave of interviews, when we asked: "Do you think there is a long-term deep effect on your life from all this?", a large majority answered in the affirmative, yet very clear relationships with background variables also emerged (LONGTERM2). Most of the "new roots" characteristics were strongly correlated with a perception of long-term personal change (young head of household, young children, smaller families, better education, homeowners with relatively fewer years of residence). Respondents with higher incomes however, were less likely to perceive deep personal change. With one exception, all of the inner-ring respondents attested to their sense of a deep change. The wording of the question in the second wave did not get at the direction of the change, so we cannot know if the deep changes were considered positive or negative. However, living through such a crisis apparently has considerable impact upon people's lives, and their perceptions of self.

During the first wave of interviews, people were reporting negative interpersonal relations as a result of the changes in their daily routines, and their feelings of strain and tension. Their need to be dependent on relatives and friends could lead to disappointments and to further strained relationships. Other people's reactions to the publicity they were receiving, to the public controversy in which they were embroiled, and to the potential for government assistance added to the strains. Table 7.5 shows the responses to a set of wave 2 questions devoted to that issue, including: "Would you say that the Love Canal situation is disrupting your normal daily routines at this point in time?" (DISRUPT2); "Do you think you have lost friends because of this?" (LOSFRND2); "Have you made new friendships as a result of this situation?" (NEWFRND2); "How are you getting along now in your family?" (FAMRELS2).

Most respondents (67%) felt their daily lives were still disrupted, months after the initial declaration by the commissioner of health. There were reports of old friends abandoning their relationships because of suspicion of the motives of Love Canal residents, or uneasiness in associating with notoriety and about one third of the respondents reported losing friends. However, many more (51%) reported making new friendships over common interests and new associations related to the crisis situation. Within families, respondents had described strains between spouses, or between parents and children in the earlier interviews, but by the time of wave 2, family relations were reported improved by 41%, and only 15% said things had gotten worse within their family. Families with young children reported more disruptions and stronger tendencies to lose friends, and to gain new ones, than did other families. (In fact, of all the background variables, having young children has emerged most frequently as an important explanatory factor in our data.)

Respondents with higher income and education reported somewhat less disruption. Again, income and education provide "coping resources" that may ease the impact of a hazardous situation. The making of new friends in the crisis situation shows strong correlations with "new roots" social characteristics— younger adults with young children, higher incomes, better educations, homeowners. Those with the longest years of residence were less likely to report making new friends.

Improved family relations also were reported by younger, better-educated homeowners. Overall, it appears that those who had better personal coping resources (income, education) and those most vulnerable (young, with young children), who were most likely to become the activists (Stone & Levine, 1985), also gained the benefits of new friendships and a pulling together of the family in facing the early stages of the Love Canal crisis.

The last set of findings are responses to a series of questions, in the second wave, on a possible sense of stigma that Love Canal people might be experiencing. We asked respondents: "What kinds of feelings do you think others have had about you as a Love Canal resident?" The interviewers asked for specifics about relatives, friends, other Niagara Falls residents, other people in general,

TABLE 7.5
Interpersonal Relations
Response Distributions and Correlations (Gamma) with Background Variables

		(1) AGE + young	(2) KID1AGE + young	(3) KIDNMBR + more	(4) INCOME + higher	(5) EDUCATN + more	(6) RESYRS + longer	(7) OWNRENT + owner	(8) RING + inner	(9) AWARLC + earlier
DISRUPT2 (n = 39)	yes—67% no—33%	.20	.35	-.20	-.25	-.20	-.12	.12	.09	.04
LOSFRND2 (n = 36)	yes—34% neutral—8% no—58%	.11	.26	-.05	.01	-.10	.28	.57	-.08	.16
NEWFRND2 (n = 37)	yes—51% neutral—8% no—41%	.81	.34	-.11	.56	.42	-.40	.24	-.13	-.06
FAMRELS2 (n = 36)	better—41% same—44% worse—15%	.34	-.19	-.11	.08	.42	-.10	.47	.04	.01

(1) AGE is age of respondent, or head of household.
(2) KID1AGE is age of youngest child in the family.
(3) KIDNMBR is the total number of children in the family.
(4) INCOME is total family income.
(5) EDUCATN is number of years of schooling of respondent, or head of household.
(6) RESYRS is number of years of residence in the Love Canal neighborhood.
(7) OWNRENT is whether respondents are homeowners or renters.
(8) RING is location of Love Canal residence, inner ring (nearest the canal) or outer ring.
(9) AWARLC is when respondent was first aware of possible danger from chemicals buried at Love Canal.

other people in the United States and people in government. The findings from these questions are summarized in Table 7.6. (Correlations are not reported for this last group of items because no clear relationships with background variables emerged.)

Love Canal residents perceived that relatively small proportions of "relevant others" held positive attitudes toward them at this point in the crisis. Friends most often were seen as having positive attitudes, but even here it was a bare majority (59%). Lowest in the positive column were other Niagara Falls residents. City government had been decidedly unsympathetic to the plight of Love Canal people, and in turn the Love Canal residents had come to realize that the city government was representing the interests, if not the sentiments, of wider public opinion. The city was not happy about the adverse publicity Love Canal was generating in an area whose economy was based on the chemical industry, and on tourist trade.

Perceptions of negative attitudes toward Love Canal residents was highest vis-á-vis government officials, and responses about the officials' attitudes also was most polarized (only one "neutral" response). Love Canal residents had been subjected to a stream of conflicting and ambiguous research reports, policy statements, assistance programs, public and private pronouncements from government officials for almost a full year at the time of the wave 2 interviews. Government assistance has continued to be one of the sources of controversy at Love Canal. As early as 6 to 7 months after the public declaration that a crisis existed, there was a clear division in perceptions between those who felt government officials were negative and those who felt they were positive in their attitudes. Thus, the final, and very clear hazard of an ambiguous situation such

TABLE 7.6
Perceived Attitudes of Others Toward Love Canal Residents

Attitudes of:	Positive	Neutral	Negative	
Relatives ($n = 31$)	52%	32%	16%	(percentages sum across)
Friends ($n = 27$)	59%	27%	15%	
Niagara Falls Residents ($n = 31$)	35%	42%	23%	
People in general ($n = 21$)	48%	52%	0	
People in the U.S. ($n = 22$)	50%	41%	9%	
People in Government ($n = 25$)	40%	4%	56%	

as Love Canal is a crisis of confidence and trust in government. This issue is discussed in great detail elsewhere (Fowlkes & Miller, 1982; Levine, 1982).

SUMMARY AND CONCLUDING COMMENTS

This study has presented data on the distribution and social background correlates of Love Canal residents' perceptions of the hazards of the toxic chemical waste dump near which they lived, and changes that the crisis made in their lives. Although the sample size was small and hastily drawn to capture residents' perceptions in the midst of a quickly developing series of crisis events, the researchers have had ample time to confirm their findings in what has turned out to be a 7-year, still-continuing research project. Discussions with researchers at sites of other manmade, invisible, technological disasters or crisis events, and with the victims of such events, confirms that the findings and our interpretations "ring true" in locations across the United States.

For the residents of Love Canal, living with "threats to people and what they value" (Kates & Kasperson, 1983) meant living in a situation of overwhelming personal concern, and being subject to decisions that affected their entire way of life. Perrow, describing Slovic, Fischhoff, and Lichtenstein's 1981 study of risk perceptions, says this about hazards people fear the most: "The dimension of dread—lack of control, high fatalities and catastrophic potential, inequitable distribution of risks and benefits, and the sense that these risks are increasing and cannot be easily reduced by technological fixes—clearly was the best predictor of perceived risk" (Perrow, 1984, p. 328).

With the exception of "high fatalities," this definition fits the situation that the Love Canal people experienced. Health, well-being of self and children, home and property as a life-long economic investment, and neighborhood as a major "social investment" were all threatened. Neither the corporation that buried the materials, nor the government authorities who permitted this to happen, nor those who were involved in dealing with the crisis as part of their professional roles were suffering the losses that the Love Canal people faced.

To take action or to do nothing were equally weighty decisions that had to be made in an ambiguous time frame, and without clear knowledge. The situation was one of great uncertainty, strain, and controversy. The questions that the Love Canal residents faced regarding threats to health, toxicity of the environment, and acceptable living conditions remain unanswered up to the present time. The issues are encompassed under a second risk-related factor described by Perrow. These risks are "unknown, unobservable, new, and delayed in their manifestation" (Perrow, 1984, p. 326).

The residents' perceptions have undoubtedly changed since 1978 and 1979. Since the time this research was done, many more events have occurred at Love Canal, as all interested parties tried to cope with this slowly unfolding disaster.

Many of the responses by government and the responsible corporation are unprecedented. Major questions about the migration of chemicals at Love Canal, effects on health, and habitability of homes purchased by the state remain unanswered. Problems may have emerged in peoples' lives that were not evident in the early months. We do not know whether effects remained stable with the passage of time. We know for example, that moving from the area created financial difficulties in the short run, but also affected respondents' perceptions of their health, and eased interpersonal tensions. The impact of the crisis did not diminish in the same way as in natural disasters; at Love Canal, as time passed, more aspects of crisis were revealed, with related impacts on people's lives. We do not know whether people became accustomed to them, and coped with new challenges in ways that improved their lives, or whether they feel diminished and emotionally wrought long after the start of the long-term crisis.

Ideally, we will have no more Love Canals. Realistically, the physical and social processes that create such disasters are in place all over our country, and in many other industrialized countries around the world. The data presented in this report should be useful in themselves, because the victims of similar disasters can be expected to suffer anxieties similar to those of the Love Canal residents. Ongoing research efforts are essential so that we may come to know the dimensions of this new problem resulting from life in industrial society, and make those dimensions a part of the background of information available to scientists, decision makers and the general public. In this way the social scientist can play an important role in ameliorating, if not preventing, the harmful social-psychological effects of our newest form of disaster.

ACKNOWLEDGMENT

We gratefully acknowledge the research assistance of Joe Heung Park.

REFERENCES

Axelrod, D. (1981). In the matter of the Love Canal chemical waste landfill site located in the city of Niagara Falls, Niagara County, State of New York: Supplemental Order 2/8/79, *Love Canal, A special report to the Governor and Legislature*, (pp. 62–67). Albany, NY: New York Department of Health.

Barton, A. (1969). *Communities in disaster*. New York: Doubleday Anchor.

Fowlkes, M. R., & Miller, P. T. (1982). *Love Canal: The social construction of disaster*. Washington, DC: Federal Emergency Management Agency.

Gibbs, L. (1982). *Love Canal: My story*. Albany, NY: SUNY Press.

Janis, I., & Mann, L. (1976). Coping with decisional conflict. *American Scientist*, 64, 657–667.

Kates, R. W., & Kasperson, J. X. (1983). Comparative risk analysis of technological hazards (a review). Washington, DC: Proceedings of the National Academy of Sciences, 80, 7027–7038.

Lang, K., & Lang, G. (1964). Collective responses to the threat of disaster. In G. Grosser, H. Wechsler, & M. Greenblatt (Eds.), *The threat of impending disaster: Contributions to the psychology of stress* (pp. 58–75). Cambridge, MA: MIT Press.

Levine, A. G. (1982). *Love Canal: Science, politics and people.* Lexington, MA: Lexington Books, D.C. Heath.

New York Department of Health (NYDOH). (1978). *Love Canal: Public health time bomb.* A special report to the governor and legislature. Albany, NY: Author.

Perrow, C. (1984). *Normal accidents: Living with high-risk technologies.* New York: Basic Books.

Quarantelli, E., & Dynes, R. (1977). Response to social crisis and disaster. In A. Inkeles, J. Coleman, & N. Smelser (Eds.), *Annual review of sociology* (Vol. 6, pp. 23–49). Palo Alto: Annual Reviews.

Short, J. F., Jr. (1984). The social fabric at risk: Toward the social transformation of risk analysis. *American Sociological Review, 49*(6): 711–725.

Slovic, P., Fischhoff, B., & Lichtenstein, S. (1981). Perceived risk: Psychological factors and social implications. *Proceedings of the Royal Society of London, A 376*, 17–34.

Stone, R. A., & Levine, A. G. (1985). Reactions to collective stress: Correlates of active citizen participation at Love Canal. In A. Wandersman & R. Hess (Eds.), *Beyond the individual: Environmental approaches and prevention* (pp. 153–177). New York: Haworth Press.

Whalen, R. P. (1978). In the matter of the Love Canal chemical waste landfill site located in the City of Niagara Falls, Niagara County: Order 8/2/78. In New York Department of Health, *Love Canal: Public health time bomb* (pp. 27–32). Albany, NY: New York Department of Health.

Author Index

Page numbers in *italics* indicate complete bibliographic information.

Subject Index

A

Agent Orange, 109
appraisal, 36, 37
  secondary, 37
asbestos, 3–15, 71, 90
  occupational exposure to, 3–15, 90
asbestosis, 4, 9

B

Bhopal, India, 89
brochogenic carcinoma, 4

C

cancer, 4, 5, 6, 7, 8, 9, 35, 50, 61, 90
  gastrointestinal, 4
  lung, 4, 9
  vaginal, 5
cardiovascular diseases, 36
chronic stress, 15, 35–45
control, 37, 40–44, 47, 50–51, 53, 54, 65,
    66, 67, 68, 114, 122, 124, 128
  locus of, 7, 47, 51, 53, 54, 65, 66, 68
  perceived, 42, 43
coping, 5, 6, 15, 36, 37, 65, 72, 79, 84, 87,
    109, 125, 128, 129
  direct action, 37
  emotion-focused, 84
  palliative, 37

D

coronary heart disease, 9, 36
crowding stress, 37

demoralization, 79, 83, 84, 85, 93, 94, 95,
    97, 98, 99, 100, 101, 102, 103, 104,
    105
denial, 5, 14, 15, 50, 65, 66, 67
depression, 36, 38, 39, 40, 42, 47, 51, 53,
    54, 55, 56, 57, 58, 59, 62, 63, 64,
    65, 66, 67, 68, 72, 94, 95
diethylstilbestrol (DES), 5
dioxin, 112
disasters, 41, 42, 44, 45, 47, 48–53, 65, 79,
    87, 105, 106, 109, 110, 128, 129
  natural, 41, 42, 44, 48, 50, 51, 79, 129
  technological, 41, 42, 44, 45, 48, 50, 51,
    52, 65, 128

E

encephalopathy, 19
environmental impact statement (EIS), 71–87
environmental stress, 64, 102

G

gamma–aminobutyric acid, 29
gastrointestinal cancer, 4